Management Principles for Safety and Occupational Health Managers
By: Fred Fanning

DEDICATION

To my paternal grandfather, Fred Eldridge Fanning.

CONTENTS

Fred E. Fanning

INTRODUCTION

Thank you for choosing this collection of chapters. I was a safety and occupational health manager for over twenty years. I also received a variety of training that was intended to improve my safety and occupational health management skills. I would like to share thoughts from my experiences as lessons for you.

In the 1980s, I was trained in Total Quality Management (TQM). TQM is a process that works for continuous improvement. I found two useful things from my training and three years of experience with this program. One is the Program Evaluation and Review Technique or PERT. This scheduling technique is well worth learning. The second thing is the wisdom that what gets measured, gets done. Management supports whatever measurements they selected and worked towards them.

In the 1990s, I was trained in Management by Objective or MBO. This program involves management developing the goals of the organization. These goals are then sent to supervisors and managers, then delegated to workers who are held accountable by the managers for completing these goals. I did not find any application for MBO to the management of safety and occupational health. Time spent on this management method adds no value to your job.

I have tried not to duplicate the things that can commonly be found in traditional management books. This book focuses on the things you probably will not hear about or be trained on. However, I think the lessons in this book are crucial to your success.

For traditional management information and techniques, I recommend reading the One Minute Manager series. I also like the early books from Blanchard and Hershey as well as a book by Carnegie. I have found them to be in line with the management techniques needed to manage safety and occupational health programs. I have listed the best books here:

- The One Minute Manager, by Kenneth Blanchard, Ph.D. and Spencer Johnson, M.D.
- Leadership and the One Minute Manager, Ken Blanchard, Ph.D., Patricia Zigarmi, and Drea Zigarmi.
- Putting the One Minute Manager to Work, Ken Blanchard, Ph.D. and Robert Lorber, Ph.D.

- Who Moved My Cheese? Spencer Johnson, M.D.
- The One Minute Manager: Build High Performing Teams by Ken Blanchard, Ph.D., Donald Carew and Eunice Parisi-Carew.
- The One Minute Manager Meets the Monkey, Ken Blanchard, Ph.D., William Oncken, Jr, Hal Burrows.
- Management of Organizational Behavior, Paul Hershey and Ken Blanchard, Ph.D.
- How to Win Friends and Influence People, Dale Carnegie.

These books provide you with the essential information you can use to develop your style of managing. It is important to be true to your style; however, if something is not working I do not mean for you to keep doing it. Do those things that work and stop doing those that do not.

I start by addressing the basics of management using behavior rather than just methods. I also believe that building knowledge and managing it properly makes your work consistent and timely. I address the need for a budget and your role in that process. In that same thread, I address risk management, which I think is critical. I use the second thread to discuss personal skills that I feel are essential to your success. I support management over leadership. In the third thread, I discuss principles and methods that many may not consider management, but I submit they are essential. The last topic I touch on is meant to help you. SOH managers can be exposed to stress that can damage their health and relationships. I encourage you to take care of yourself and those you care for.

I hope you enjoy the information in this book. It is meant to save time and money for safety and occupational health managers who are trying to do the best job possible in preventing accidents. For more information on this and other safety and occupational health topics, please visit the bibliography at the end of this book.

CHAPTER 1 –THE ABCS OF MANAGEMENT

Introduction

The Safety and Occupational Health (SOH) manager gets work done through direct reports primarily because they must if they want to get paid. There are also times when a SOH manager must get work done through others whose pay they do not affect. With these people, the SOH manager must use influence. In either case, it helps to know a little bit about why people behave the way they do so they can get work done by them.

First, it is important to recognize that work groups succeed or fail to get work done because of employee performance. For the most part, good performance creates successful work results, and poor performance creates failed work results. So, what is performance? Performance is the combined effects of employee behaviors. SOH managers must be familiar with the *ABC*s of managing and how to improve *B*ehavior and the resulting performance.

You do not need a degree in Psychology to understand why employees behave the way they do. Employees have *A*ntecedents that include beliefs and values. These *A*ntecedents trigger *B*ehaviors or human actions that result in *C*onsequences, or actions taken in response to *B*ehaviors that either reinforce or punish the *B*ehavior. That is what makes up the *ABC*s of management.

Antecedents

*A*ntecedents or beliefs and values are inside a person's head. They cannot be seen or measured, and that makes them very hard to change. They are learned throughout an individual's childhood, adolescence, and adulthood. In many cases, they are a combination of what is learned and how the individual synthesizes or integrates the information into their mind. These beliefs and values impact the employee's *B*ehavior. Some examples might include the belief that "accidents will not happen to me", "speed limits do not apply to me", "safety is someone else's responsibility", or "it is not my job".

3

Behavior

In contrast to *Antecedents*, most *Behaviors* are observable and therefore can be measured and managed. *Behavior* is action and includes speaking, acting, and performing physical functions. *Behaviors* might include reading a reference book, writing a report, placing on the emergency brake in the car when parking on a hill, or sealing boxes on an assembly line.

Consequences

Consequences are the events that follow *Behaviors*. *Consequences* increase or decrease the probability that the *Behaviors* occur again. *Consequences* might include receiving a ticket for speeding, sliding off the road when driving too fast for road conditions, or losing a job because the boss did not think the employee was a team player. The amount of influence *Consequences* have on *Behaviors* depends on individual perceptions of whether the *Consequence* is positive or negative, large or small, or personal or public. There are also three types of *Consequences* that include reinforcement, punishment, and extinction.

Reinforcement can be positive or negative. Positive reinforcement can be defined as "do this, and you will be rewarded." Examples of positive reinforcement are: a SOH manager congratulates an employee for making a safety suggestion, or a SOH manager thanks an employee in front of her peers for completing a report on time. The opposite of positive reinforcement is negative reinforcement. Negative Reinforcement can be defined as "do this or else you will be penalized." Examples of negative reinforcement are: a SOH manager tells an employee if they submit a report on time they will not have their performance docked or a SOH manager tells an employee that if they come to work on time they can receive pay for the entire day.

Punishment is defined as "if you do this, you will be penalized." Examples of punishment include a SOH manager threatening an employee with termination if the employee fails to show up for work or a SOH manager terminating an employee for failing a drug test.

Extinction is defined as "ignore it, and it will go away." Examples of extinction are a new employee who is whining about the level of work he has to complete each night, while the SOH manager lets the employee whine so he will learn he does not get a response from the supervisor each time he whines. Soon, the employee stops.

Summary

SOH managers can influence employee Behavior by controlling the Consequences the employee receives for that Behavior. If a SOH manager wants to encourage Behavior demonstrated by an employee, he or she can use positive or negative reinforcement, thereby encouraging the employee to keep up the good work. A SOH manager can also punish an employee for Behavior he was told not to perform. By using all four types of Consequences in response to Behavior, the SOH manager can cause an employee to continue a particular type of Behavior while stopping unwanted Behavior. Studies have shown that positive reinforcement provides long-term improvements in Behavior while negative reinforcement, punishment, and extinction provide an immediate improvement that usually drops off with time. To be effective, SOH managers must respond to Behavior in a way that improves performance over the long term. It can be as easy as *A, B, C.*

CHAPTER 2 – RESOURCE MANAGEMENT

Introduction

An area of consistent weakness I have seen in working with Safety and Occupational Health (SOH) Managers over the years is their lack of understanding of resources for the SOH programs and office. If a SOH manager does not understand what it takes to run the programs and the office, he or she is left to the mercy of others to provide money. Over the years, I have found when we leave ourselves open to the mercy of others, we do not get what we need. Without a budget, it is hard to plan and even more difficult to maintain a level of spending to keep the programs and office operational. Try hard to get authority for a budget at your level. If possible, prevent anyone from putting the SOH programs and office under the administrative budget that many offices share.

It is essential for SOH managers to learn the basics of budgeting. If there is not a budget for the SOH office and programs, you need to create one. Work tirelessly until this is accomplished. The foundation or justification for a safety budget is "Safety reduces the costs of work-related injuries, illnesses and property damage to the organization." This needs to be used when requesting money for the budget. SOH Managers must have some idea of how many injuries and illnesses as well as how much property damage they either want to prevent or have already avoided. SOH does not make money for the organization. Therefore, it is essential to create an offset to the budget you request. A good example is a SOH manager requesting an $85,000.00 budget to reduce the cost of illnesses, injuries, and property damage by $186,000.00. Always use common sense; you cannot ask for more money than you can save the organization. Typically, a budget is broken down into broad categories that include human resources, training, travel, awareness material, incentive awards, and hazard corrections. The biggest part of any budget is the cost of labor. Let's tackle that item first.

Human Resources

Your goal here is to determine how many personnel are needed to implement and maintain the SOH program and operate an office in the organization. To determine this, first identify all the work that needs to be done in a typical year. This might include conducting inspections, investigating accidents, recommending personal protective equipment, training employees, and tracking hazardous material, just to name a few. Second, you need to identify how many hours of work each task takes. Total the hours for all tasks and divide by 2,080. The answer is how many personnel are needed. Why 2,080? This is the number of hours the typical person works on a full-time job in a year. For example, if the hours add up to 20,800 divide that by 2,080, and the answer is 10. It will take ten full-time employees to do the work you have identified; however, you may not get ten full-time personnel.

The next step is to determine if this work can be done by full-time employees, collateral duty SOH representatives, or a mixture of both. It is up to management to determine how this program is operated; however, the SOH manager is expected to provide a recommendation. If the organization chooses to use only collateral duty SOH representatives the time they spend on collateral duties must equal the number of hours needed. Some organizations use a ratio of one collateral duty SOH representative for five hundred employees. This ratio works for a collateral duty SOH representative that spends at least 8 hours per week on safety duties. The more time an individual spends on safety, the fewer personnel you need. In contrast the less time an individual spends on safety, the more personnel the organization needs. This can be modified so that a collateral duty SOH representative is assigned from different sections. For example, an organization has three shifts with three lines. Even though there are only three hundred workers, it would benefit the organization to use three collateral duty persons instead of the one determined by the ratio. The ratio is simply a starting point and should be modified to meet your organization's needs. The organization's leadership may choose to use only full-time professionals. In this case, the individual has a full forty-hour week to do safety work and can do a great deal more than a collateral duty SOH representative. It is common to have a full-time SOH professional for each plant or organization. This ratio is done without much regard for the number of employees. Another approach is to have a full-time SOH professional for every 3,500 employees if you prefer a ratio to employee count. Also, the organization's leaders may decide to use a mixture of both collateral and full-time SOH personnel. In this case, it seems to work best if there is a full-time SOH professional at the headquarters level with collateral duty personnel assigned at a ratio of one collateral duty SOH representative for every one thousand employees or one per line. I do not recommend using a ratio. I always use the actual number of hours to complete tasks.

There is a charge or cost for each hour of SOH work. The full-time SOH professional costs the actual salary and benefits for that position. To calculate the collateral duty SOH representative costs, you must take the actual salary and benefits of a worker for the actual hours they perform SOH work. Salaries include the number of collateral and full-time SOH personnel that are being used. You can obtain the salary and cost of benefits from the Human Resources Office. Submit this cost in the budget.

Training

There must be a training program that is funded to make sure all personnel within the organization are adequately trained to do their jobs safely. This money often remains with the human resources department who allocates it as requested; however, the SOH manager should be providing input to the training based on organizational requirements as well as the applicability of the course. Input ensures that the appropriate personnel receives the proper training, saving the organization's money and effort. The SOH manager must identify all required training in a prioritized list. Submit the cost for this training in the budget.

Travel

Travel usually involves the expenses for personnel to travel to attend training as well as sending personnel to conferences and trade shows. Both are critical to benchmarking the safety program with other successful and assertive programs. Networking at these conferences and shows can provide personnel with resources that they can call upon in the future to aid the organization, usually for free. Also, new products and services can be seen that your organization may need in the future and may not know about, if not for this opportunity. Submit this cost with the budget.

Awareness Materials

Organizations need awareness materials. These are the material used in the form of posters, brochures, handouts, buttons, etc., to get the word out about hazards within the work areas, and measures that can be taken to prevent accidents from occurring or lessen the severity if they do occur. There should be approximately fifteen cents spent annually for each employee within the organization. This is a formula that works very well. If you have a high hazard organization, this amount should be fifty cents per employee. The key is to spend the money effectively on the hazards that are affecting your processes and personnel. It is also important to follow up with awareness material used to ensure it is well received by employees within the organization. If you have non-native English speaking employees, it is essential to provide some awareness material in the language they speak naturally. My experience proves that providing material in the language a person speaks not only helps them know more about the safety program, but also gives them some incentive to become an active supporter of safety. There is also a need to look at the different age groups of employees. Younger employees seem to like active, busy posters and awareness material while older employees appear to like straightforward single message material. Do not forget about visual and hearing impaired employees. Focus your material to a broad audience to reach all or most of your workforce.

Incentive Awards

There are many discussions and even arguing about the effectiveness of incentive awards programs. In over twenty years of safety experience, I have found that these programs can work if done right. They cannot be handled haphazardly or without clear intent. This program should consist of incentive awards and earned awards. Both programs complement each other and together form a solid program that keeps the SOH program positive while gaining employee support. The incentive awards should be used to garner support for the program. These should be low-cost items that can be given to employees when they do something right. Too many times we tend to catch employees doing things wrong. This process allows management to catch them doing the opposite. The second part should be an awards program that requires the individual to earn the award through some defined criteria. These awards can be for working so many hours without an accident or driving so many miles without an accident. The criteria can also be for lost time accidents. Criteria allow the organization to maintain an active accident reporting program. These awards must be kept separate and not given out for small or minimal effort. The incentive awards are typically kept to $5.00 or less while earned awards should come in five different levels that have a progression for an employee to strive toward. Furthermore, the award should be provided to an employee in a manner they feel comfortable. There are some employees who do not wish to receive an award before a big group. Do not embarrass these employees. Award their items in front of their section. Include all personnel receiving an award in newsletters or notices placed on company bulletin boards.

Hazard Correction

Hazard correction is an area that demonstrates to the workforce that management takes safety seriously and is not afraid to put their money where their mouth is. Identify hazards through inspections and reports of unsafe or unhealthful working conditions. After identifying hazards, there should be a risk assessment code assigned to each hazard. Hazards are then prioritized based on the risk assessment code so money is spent on the hazards most likely to cause an accident that will result in injury or property damage. See chapter 5 for a matrix that can be used to assign the risk assessment code. Correct all high hazards before moderate hazards, and tackle moderate hazards before the low. If the safety program for your organization is to be successful, there must be some money applied to it. Base the amount of money on the size and complexity of the organization. The resources must be clearly identified, and money or manpower assigned to them. This money does not have to go to the SOH manager, but rather it needs to go to the individual who corrects the hazards. This could fall under facilities, human resources, or operations. The SOH manager must act as the steward of these requests since he or she corrects safety hazards. These budget items could be in the hundreds of thousands of dollars.

Summary

There are two kinds of managers in an organization: those with a budget and those without one. Those with a budget plan the work and then execute it per the budget. Those without a budget plan and then spend months looking for money that probably does not support the plan they developed. The four most valuable tips for budgeting are:

- Know the basics of budgeting.
- If there is not a budget, create one.
- Do not ask for more money than you can save the organization.
- Execute the approved budget.
- Plan for a budget and get the results.

CHAPTER 3 – HUMAN RESOURCES MANAGEMENT

Introduction

Human Resources can be the most expensive cost for an organization. Management plays a significant role in selecting and hiring employees and needs to manage such an expensive resource better. This is done by leveraging human resources management principles. These principles consist of hiring the right person, ensuring potential employees are technically competent and physically capable, bringing new employees on board correctly, and managing aspects of the new employee's career with the organization.

Supervisor's Role

The supervisor plays a central role in the hiring process. Supervisors must make an up-front investment by taking the time to consider critically and then effectively communicate to the Human Resources Office the critical knowledge, skills, and abilities (KSA) required by a job as identified by a job analysis. KSAs begin with a thoughtful conversation in which the supervisor furnishes information that enables a human resources specialist to design a job more efficiently, market the vacancy better, and assess applicants more thoroughly. The supervisor again plays a vital role when conducting structured interviews as the last step in the assessment process. The supervisor then makes the hiring decision. Later the supervisor must ensure the new employee is brought into the new job properly to ensure he or she knows about required policies and procedures, use of equipment, reporting hazards and accidents, and how to identify training and growth needs.

Human Resources Specialist's Role

The human resources specialist plays a supporting role in the hiring process. They participate with supervisors in the up-front investment by taking the time to consider critically and then ensure they understand the critical knowledge, skills, and abilities required by the job. It is at this point that the human resources specialist takes the information provided by the supervisor that was outlined in the previous chapter and designs the job, markets the vacancy and assesses applicants.

Safety and Occupational Health Specialist's Role

The Safety and Occupational Health (SOH) specialist also plays a supporting role in the hiring process. They provide information about hazards associated with a job, extreme weather exposure, ergonomic issues, and any special safety skills that an employee may need. The SHE Specialist works with the human resource specialist to develop a job hazard analysis.

Job Analysis

Job analysis is the foundation of recruiting human resources and is vital to selecting the proper employee. Identifying the best person for the job requires that the supervisor fully understands essential duties of the job and the environment in which an employee performs the job. Through the job analysis, the supervisor systematically identifies the knowledge, skills, and abilities necessary for success on the job. Then the supervisor, with the help of a human resources specialist, develops valid and efficient selection tools.

The supervisor need not conduct a job analysis every time he or she wants to fill a job vacancy; however, it is essential the first time a job is created and filled. If job openings within the same occupation frequently occur, the supervisor may rely on selection tools developed from recent job analyses of that occupation. These selection tools include structured interviews, written tests, use of an assessment center, and work samples.

Whether a job analysis should be conducted for a job depends on whether the job is new, and if not new, how current the existing job analysis is. Periodic review of job analyses is critical. If the requirements of the job frequently change, the supervisor should consider the job analysis before each job being filled to ensure the selection tools are still valid. For a job that changes very little, by contrast, the job analysis may need to be reviewed less frequently; perhaps annually regardless of the number of times you fill the job.

Job Hazard Analysis

A SOH professional conducts a job hazard analysis. The job hazard analysis focuses on job tasks to identify hazards before they occur. In contrast, a job analysis covers a much broader scope of what the job entails. Like the job analysis, the job hazard analysis includes the relationships that exist between the worker, task, tool, and the work environment, but does not focus on it. The job analysis must precede the job hazard analysis because it provides the tasks of the job. The job hazard analysis facilitates the identification and control of hazards associated with the job.

It is important to identify the physical demands of the job up front. These include walking, standing, climbing stairs and ladders, and lifting. Noise and visual effort are also necessary. Another important piece of information is how frequent the requirements are experienced. Note the frequency as sustained, intermittent, or seldom. The results of this information determine the requirements for a pre-employment physical. Physicals are an important part of any hiring process. Without a proper pre-employment physical, the employer accepts responsibility for the employee as is. Identifying previous injuries could result in significant cost savings if an employee's previous condition is exacerbated by the new job.

Identifying these requirements involves noting the environment in which the job is performed. This includes heat, cold, height, underground, in darkness, and inclement weather. This information allows the supervisor also to identify protective equipment and clothing to reduce the risk to the employee from the specific environmental requirements. This information also supports the pre-employment physical in such areas as cold and hot weather injuries where a previous injury may leave an employee susceptible to future injuries.

Classification

The next step after the job hazard analysis is to have the job classified. This is called Position Classification and describes the process through which jobs are assigned to a pay system, series, title, and pay grade, based on a consistent application of position classification standards. Positions are classified to achieve uniformity and equity using a standard reference across organizations, locations, and agencies. Classification rules cover one or more occupations, usually including a description of the work performed; official titles; and criteria for determining pay grades. Most human resources offices have developed grading guidance, broad standards that serve as practical guides, and provide criteria for determining the pay level of work. These can typically be obtained from the human resources office serving the supervisor's organization. Position classification standards and guidance also distinguish between white collar and blue collar (trades, craft, and labor) jobs. It is easy to confuse classification with qualifications; however, each has its distinct purpose.

Classification pertains to a job and the evaluation process that determines the appropriate pay system, occupational series, title, and grade. Qualifications pertain to a person and describes the knowledge, skills, and abilities a person must have to be successful. Normally there are three kinds of competencies:

- Required Competencies are those required for the position.
- Enabling Competencies are not required by the position but are instrumental in assisting the incumbent employee to perform the job successfully. An incumbent should work to develop these competencies.
- Developmental Competencies are not required by the position, but necessary for the incumbent to move up to the next pay grade.

Job Description

A job description is a statement of the principal duties, responsibilities, and supervisory relationships of the position. Simply put, a job description indicates the work to be performed by the employee incumbent to the position. The purpose of the job description is to document the major duties and responsibilities of the position. The job description cannot spell out every possible activity during the work day. Human resources offices often maintain a library of job descriptions.

Advertising the Vacancy

Once the job description has been completed, a vacancy advertisement must be written. This is the description that people will see that will encourage them to apply for the job. The job should be advertised in professional associations boards, internet job boards, local help wanted ads and word of mouth. Get the word out with the goal of getting a pool of applicants from which to choose, selecting those who will be interviewed, and eventually one that will be hired.

Employee On-Boarding

Once the employee is hired, he or she must be brought on board properly. After you take all that time to hire the right person you do not want them to leave. A formal onboarding process is necessary.

Employees must complete the proper paperwork for insurance, taxes, retirement, pay, emergency procedures, lunch and break times, work schedule and times, and receive an introduction to the new organization. After that, the employee is sent to the actual office where he or she is introduced to other employees. A workstation should be ready for this employee to go to work immediately. If the worker gets a computer, the supervisor should convey the temporary passwords and appliances ready to go. If a phone is provided, it must be assigned the proper number and voice mail must be ready to set-up. This is usually followed by a tour of the facility and introductions to other key employees. Do not forget to explain locations for lunch, emergencies and how friends and family can contact the new employee.

Career Management

Once the employee is properly onboard, it is important to provide them with information about training opportunities and mentoring that is offered by the organization. The employee's supervisor should explain how the employee can prepare for promotions and upward mobility within the organization. Once a good employee is hired it is profitable to keep them on the payroll if possible. The employee will probably look for a promotion and more money after their first year on the job. To keep the employee, the supervisor can help them get promoted within the organization rather than lose them to another organization.

Summary

With SOH managers playing such an important role in the work life of Americans, it is important to use all the tools available to hire the best employees. When you hire, a SOH specialist using this information can help you get it right. Once you get the right person, take care of them and assist them in growing in their career. You do not have to do this alone. You should work with the Human Resources Staff of your organization to get it right.

CHAPTER 4 – RISK MANAGEMENT

Introduction

Safety and Occupational Health (SOH) managers need to have a way to focus their efforts to correct hazards. Very seldom is there all the money needed, so decisions are made about what to correct now and what to correct later. One way to do that is by correcting hazards based on the risk they pose. This is the best method I have found, and it is easily explained to managers and budget staff. After hazards are identified and before any attempt is made to correct them, determine the risk they pose. Once that is done, focus on correcting the worst (highest hazard) first and then moderate and low hazards as money permits. In this way, money is spent to repair the most dangerous first, thereby preventing the more serious accidents, illnesses, and property damage.

Risk Analysis

An analysis of all hazards must be made to determine the degree of risk. Annotate the results of this analysis in an inspection log. Assess hazard risks regarding hazard severity using Table 1 and accident probability using Table 2. From those tables, a risk assessment code (RAC) can be determined using Table 3.

To make this easier to understand, let me outline a few instructions. First, identify the hazard severity category as described in Table 1 somewhere between a Category I and IV. This should be if an accident or exposure occurs that leads to an injury or illness. Pick the most common severity category that you believe might happen.

Next, identify the category of probability between A through E that an accident or exposure will occur using the definitions in Table 2. Again, pick the most common probability level, not the most extreme.

Category: I
Description: Catastrophic
Definition: Loss of ability to accomplish the duties of organization or failure of duties by the organization. Death or permanent total disability. Loss of major or critical system or equipment. Major property damage.

Category: II
Description: CRITICAL
Definition: Significantly (severely) degraded capability to conduct duties of the organization. Permanent partial disability, temporary total disability exceeding three months. Extensive (major) damage to equipment or systems.

Category: III
Description: MARGINAL
Definition: Degraded capability of the organization to perform duties. Minor damage to equipment or systems, or the environment. Lost days due to injury or illness are not exceeding three months. Minor damage to property.

Category: IV
Description: NEGLIGIBLE
Definition: Little or no adverse impact on the organization to perform duties. First aid or minor medical treatment. Slight equipment or system damage. Little or no property damage.

Table 1 - Hazard Severity

Category: A (Frequent) - occurs very often, continuously experienced. Occurs very often in equipment service life. Is expected to occur several times over the duration of a specific operation. Occurs very often in a person's career

Category B (Likely) occurs several times. Occurs several times in equipment service life. Is expected to occur during an operation. Occurs several times in person's career.

Category C (Occasional) occurs sporadically (irregularly, sparsely, or sometimes). Occurs some time in equipment service life. May occur about as often as not during an operation. Occurs several times in service life. Occurs some time in a person's career.

Category D (Seldom) could occur at some time. Occurs in equipment service life, but only remotely possible. Not expected to occur during a particular operation. Occurs as an isolated incident during a person's career.

Category E (Unlikely) can assume will not occur, but not impossible. Occurrence will almost never occur in equipment service life. Can assume will not occur during a specific operation. Occurrence not impossible, but may assume will not occur in a person's career.

Table 2 - Accident Probability

Go to table three and identify the hazard severity category from table 1 (I, II, III, IV), put your finger on it and run your finger across the columns until you have the probability level (A, B, C, D, E) you selected. Now you have the number that compares severity and probability. If your hazard severity category was II and your probability level was B, the number would be 2. So, your RAC would be II-B-2, which would be an RAC of 2.

Accident Probability					
Hazard Severity	A	B	C	D	E
I	1	1	2	3	5
II	1	2	3	4	5
III	2	3	4	5	5
IV	3	4	5	5	5

Table 3 - Risk Assessment Code Matrix

This means that the accident or exposure would cause a critical situation that would significantly degrade the organization's capability to conduct its mission, resulting in a permanent or partial disability, or temporary total disability exceeding three months. It would also cause extensive (major) damage to equipment or systems with the likelihood that the event would occur several times in an operation during the service life of a tool or item and is expected to occur during a specific operation. This event occurs at a high rate but intermittently (regular intervals, often). Typically, an employee could experience the event several times in a career, and it is expected to occur during a specific operation. Exposure to all employees is at a high rate but experienced intermittently.

Risk Management

You manage these risks by implementing ways to control or eliminating the risk. Do this by identifying control measures. There is a hierarchy of controls. That begins with elimination and runs through using personal protective equipment. Base this hierarchy on effectiveness. Eliminating the hazards is the most effective means to prevent injury, illness, and property damage. While using personal protective equipment is less efficient because it relies on the worker to properly use the equipment whenever it is required, this often does not occur. The hierarchy of controls is:

- Eliminate the hazard by physically removing it.
- Substitute less hazardous procedures or materials for the dangerous ones.
- Use engineering controls to isolate the worker from the hazard.
- Implement administrative controls for the hazard by changing the way people work.
- Provide and require employees to use personal protective equipment to protect them from the hazards.

Residual Risk

You can now go back through the log and identify the hazard severity category as described in Table 1 after a control measure is implemented. This should lower the category. Next, go back and identify the probability level that an accident or exposure will occur using the definitions in Table 2 now with the control measure in place. Now proceed to Table 3 and identify the hazard severity category from the new category from Table 1 and new probability level from Table 2. This number should be lower than the previous RAC selected. This new number becomes the Residual RAC and represents the risk associated with the residual hazards after control measures are implemented.

Summary

In today's fiscal environment, it is almost impossible to get all the money we need to correct hazards. It is, therefore, important for SOH managers to identify how to target corrective measures to get the best use of resources. The best way I have found is through the identification of risk that hazards pose. This helps the SOH manager identify which hazards to fix first. It also provides a method to explain why hazards are chosen. This gets money spent on corrections more efficiently.

CHAPTER 5 –LEADERSHIP

Introduction

If you need to get work done by people that do not report to you, you need to be a leader. It is important to not only understand management, but to be familiar with leadership, so you can get the work done.

Leadership Definitions

What is leadership? Dictionary.com (Leadership, 2013) defines it as a position or office of a leader, capacity or ability to lead, a group of leaders, or guidance and direction. That definition does not appear to be very helpful. Perhaps it would help to find the definition of a leader. Unfortunately, Dictionary.com (Leader, 2013) defines a leader as "a person or thing that leads." This might be the root of the problem: no agreed upon definition of leader or leadership that can be used to facilitate discussion is easily found. I prefer Don Clark's definition of leadership. He says that leadership is "the process of influencing people while operating to meet organizational requirements and improving the organization through change (The Art and Science of Leadership, 2013). The next question is "What do leaders do?" Ms. Rosalyn Carter, the former first lady, explains that "Leaders take people where they want to go, and great leaders take people where they do not necessarily want to go, but ought to be" (ArcaMax, 2013). With those two definitions in mind, I would like to address the three core principles underlying the mastery of skillful action by a leader as outlined by Kevin Cashman in his book Leadership from the Inside Out.

"Leaders must face all aspects of themselves honestly because any characteristics that they deny have underdeveloped, or cannot be seen by the leader are seen clearly by those around the leader" (Cashman, 2001). I took from this description that truth and honesty were two watchwords of a leader. Kevin Cashman further explained that "leaders must also accept responsibility for their actions and not blame others or make excuses. Leaders who make excuses are dismissed by those they lead" (Cashman, 2001). I once worked for a leader who, when he would come from meetings with his boss, directed new tasks to do with too little time and too few resources. When the employees pushed back to the leader, he often laid the blame on his boss, replying, "I tried to get us more time, but the old man would not listen to reason." I also worked for a leader that took credit when work was done well but blamed her direct reports when work went badly. In one case, she told her boss, "I know, this did not work out as I had planned either, but I had a talk with Mary about her failure on this project." This obviously lets the boss know it was Mary who was the problem. In both cases, the leader's actions harmed the employees while demonstrating poor leadership.

Be Yourself

Leaders must be themselves and not try to be someone they think others want them to be. Leaders must get work done and build a relationship with those they lead (Cashman, 2001). It takes both to be successful. Kevin Cashman did not "mean that the leader should be friends with those he or she leads, but rather should develop a professional relationship. By taking a little risk and letting the employees know who they are leaders begin to develop a relationship that makes the employees want a leader to succeed" (Cashman, 2001). Too often today, I work with leaders who want to be friends with their employees. I think Kevin Cashman is correct to point out that what is called for here is a professional relationship, not friendship. It is also important for leaders to understand why they do things. I have experienced the rush project that comes in, and the leader overreacts and pulls a team together to accomplish the task when what was needed was to give the regular staff the task with the deadline within which they could have done the work. I have often heard the saying "we always have time to do it over, but never time to do it right in the first place." I do not know whom to credit for this saying, or I would thank them personally. Why did the leader over react? Was it because of a previous assignment that went bad that cost them a job, a fear of failure, or was it a misunderstanding of the situation? In my case, the leader had a fear of failure that caused him always to overreact. Leaders must learn why he or she reacts the way they do and change their behavior when necessary.

Create Value

The third focus is that leaders create value (Cashman, 2001). This is very different than getting results. Today most leaders want to get results, but at what price? How can we know if we are creating value? He or she should ask themselves three questions. First, am I enriching life or depleting life? Second, am I opening possibilities or shutting them down? Third, am I focused on the "why" of the task rather than the "what"? Using these answers, leaders determine if they are adding value.

Summary

Safety and Occupational Health Managers that want to be leaders should use Don Clark's definition of leadership: "The process of influencing people while operating to meet organizational requirements and improving the organization through change" (The Art and Science of Leadership, 2013). They should also use Ms. Carter's explanation of what leaders do: "Leaders take people where they want to go." If we take those definitions and work to implement Kevin Cashman's three core principles, we can all be good leaders who are honest and truthful, while being ourselves and adding value for everyone involved.

CHAPTER 6 – PUBLIC SPEAKING

Introduction

To improve our ability as Safety and Occupational Health (SOH) managers, we must be exposed to a variety of methods and styles of public speaking. Through each exposure, we learn a little more and enhance our speaking ability. It is for that reason that as a member of my local Toastmasters International Club, I experienced the training methods used to make members better public speakers firsthand. "Dr. Ralph C. Smedley launched the first Toastmasters Club back in 1924" (Smedley, 1988). Dr. Smedley founded an educational organization to help its members improve communication and leadership skills. "The education program is the heart of the Toastmasters Club. It provides members with a proven curriculum that develops communication and leadership skills one step at time" (TI Education, 2009). I learned a great deal from this method and believed that all SOH managers could benefit from learning more about the Toastmasters' International Training Method firsthand by participating in the process.

Curriculum

The Toastmasters International Education curriculum focuses on the adult learner. Back in 1924 no one was focused on the adult learner or adapting teaching methods to their needs. That is, except Dr. Ralph Smedley. Dr. Smedley developed a club that allows its members to practice public speaking, listening, and critical evaluation techniques in a comfortable atmosphere where everyone is equal. He did this for three splendid reasons.

- The first result of speech training is self-discovery.
- Real communication is impossible without listening.
- We gain skill by practice, and we improve by heeding our evaluators (Smedley, 1988).

As SOH managers, we know that adults have lots of knowledge about a subject before they ever set foot in the classroom. How can we use that experience and knowledge to help us connect with the material we want to learn? Toastmasters have developed an excellent technique in the club. The club meets periodically, and visitors are encouraged to attend. When a visitor attends, he or she immerses in the club activities of speaking, listening and evaluating. They see firsthand the learning process in practice. They are then asked to join, and if they choose to, they become a member. As a new member, they are afforded the opportunity of a mentor to help them with the first task of public speaking. That task is to decide to complete the Competent Communicator manual. This manual breaks down into ten speeches that give the newcomer as well as the old hand an opportunity to practice a variety of speeches. The manual provides an explanation of each speech and learning objectives. The first speech is called the "Ice Breaker." It is simply the new member speaking about them.

Listening

What on earth has listening got to do with speaking? A significant part of public speaking is the ability to speak clearly and correctly. To do that, we must listen to how words are used, voices adjust, and how often filler words or mispronunciations occur. This is not to grade the speaker, but rather for the listener to hear how others speak and identify those same weaknesses or strengths in him or herself. New words and even new meanings are learned through listening. Also, the listener can learn how other member's speeches are delivered and speak in the same fashion when it is his or her turn. Listening is also critical to proper evaluation, and in the club method everyone has an evaluator to provide feedback.

Speeches

There is no formal instruction in a Toastmasters Meeting. Instead, members evaluate one another's presentations (How Does It Work, 2009). "In Toastmasters, feedback is called evaluation, and it is the heart of the Toastmasters educational program" (Effective Evaluation, 2009). An evaluator is assigned to listen and critique each speech. Each speech is given an oral critique in front of the whole club and then a more detailed written critique in the member's manual. There is also a grammarian, Ah Counter (someone who counts how many times a speaker says the word ah), and a timer that provides the speaker more feedback about the speech.

Feedback and Evaluation

The speaker gets a lot of feedback on his or her presentation and a chance to improve and speak again shortly. All critiques are done with the best interest of the speaker in mind. No one is there to demean or belittle a speaker, but rather to build them up. No one tells a member what to speak on, and no one passes or fails a speaker, yet improvement occurs. That is because all members are there for the same reason: to improve their public speaking skills. This works because of the club and cannot be learned in isolation (Communication Track, 2009).

Each member takes a turn at evaluation so that he or she can learn the skills of effective listening, developing feedback, and giving feedback. These skills are valuable life lessons in addition to good speaking lessons. The evaluator provides an honest reaction in a constructive manner for the speaker (Effective Evaluation, 2006). To help the evaluator, Toastmasters International provides evaluation guides, but the evaluation is clearly the evaluator's opinion and nothing more.

All speakers need feedback about their attempts to speak efficiently. Without it, speakers are speaking with blinders on. The evaluator gives the speaker the information he or she needs to improve their speaking ability. It is still up to the speaker to choose what advice to take. Multiple speech evaluations come in handy. The speaker hears from a different evaluator after each speech and from these sessions he or she starts to see a picture of what the audience sees and hears. Feedback is valuable information.

Summary

I think every SOH manager should be familiar with the Toastmasters method. I also recommend SOH managers give Toastmasters a try for one year while they complete the Competent Communicator program.

CHAPTER 7 –NEGOTIATING

Introduction

Many Safety and Occupational Health (SOH) managers come through the ranks of production, sales, and finance. It seems that no matter where a SOH manager comes from, they each fall back on the strategy that got them to this point in their career when stressed. Those from the production floor tend to focus on the production processes while former salespeople tend to concentrate on the sales processes. This method of coping with stress makes sense. If your focus got you into the ranks of management, it should be good enough to keep you there, should it not? My answer to that question is no. There are many skills that are brought from the workforce to management, but then there are other skills a SOH manager must learn to be successful. One of those skills is to negotiate. If a SOH manager is not familiar with the methods of negotiation, he or she is likely to become stressed and fall back on methods they depend on that may not be successful in a negotiating situation. However, "when you develop negotiating skills it is easier to be more confident, assertive, motivated and achieve better working habits at home and work as well" (Ortiz, 2012).

The New Oxford American Dictionary defines the word "Negotiate" as trying to reach an agreement or compromise by discussion with others. It also defines the word "Negotiation" as discussion aimed at reaching an agreement (Oxford, 2012). Most of us have negotiated the purchase of a home or car, and if you are a parent, you know what negotiating is about. Professional negotiations are very similar, but often more challenging. Negotiating usually has two opposing sides who are trying to achieve agreement so that each party gets most of what they want. Professionally, you could negotiate a real estate purchase, labor relations contract, insurance purchase, sale of goods or services, and many other acquisitions. Each party communicates their talking points and then go back and forth discussing the pros and cons. Each side gives and takes until both sides feel they have gotten all they can from the negotiations. At this point, an agreement is reached, or both sides agree that they are at an impasse. If an agreement is reached, the document is usually drafted with the agreed-upon points, and both sides' sign the document. If a standstill is achieved, the negotiations can be canceled, or a third party can be brought in to resolve the stalemate.

I do not want to make it sound like all negotiating is alike, because it is not. "Every scenario must be treated differently. In every negotiation, there is no one-size-fits-all solution for every situation" (Tideas, 2013). I believe that the process of each negotiation is very similar and is adapted to each negotiation. What I am writing here is general in nature, and from my experience, represents negotiation with people from the Western Hemisphere. Negotiating with people from other cultures and societies requires adaptation to the norms of that society.

Preparing to Negotiate

There are things that you can do before, during, and after a negotiation that can make a big difference in how successful you are. Margaret Ortiz recommends "going into a negotiation with good body language and a gentle tone of voice" (Ortiz, 2012). You should approach every negotiation with a positive attitude and pleasant demeanor. "Negotiations should be done between two people who do not plan on taking advantage of each other or demanding to win. You should always be constructive and transparent and not manipulative" (Tideas, 2013). Confrontational negotiations usually result in an impasse or a single happy winner and a sore loser. Neither of which is suitable for you. In addition to beginning with a great attitude, you must know what you need out of the negotiations.

You need to identify exactly what you want or need from the negotiations. It might seem funny, but some people go into negotiations without this idea and end up with something that does not fit their need. Stephen Covey said it best, "begin with the end in mind" (Covey, 1989). Knowing this, you can move on to learning more about what you need. You need to identify what constitutes the least acceptable terms of the agreement that you can accept. From there, determine the outcome you would most like to see. You also need to identify your goal for the negotiations. Lastly, you need to identify what the most likely outcome would be if the negotiations were not held.

Ben Robinson tells us "Information is the lifeblood of negotiation. The more you know about your position and the other side's position, the better." To be successful, you must learn as much as you can about your job before you go into negotiations. If we take a real estate lease, for example, you want to do some analysis to identify exactly what you need from this lease. This would include identifying how many square feet of space, how it is to be built-out and what type of furniture do you have or will be purchasing, whether this is a new lease or if this will be replacing a leased facility you already have. Now contrast that with the leased property. Learn as much as you can about the property that would likely include rent, utilities, taxes, maintenance, security, and cleaning costs. Ask other people in the building about it. Find out about the real estate agent you are negotiating with. How does he or she like to negotiate? The more information you can gather, the better prepared you can be to agree to tradeoffs to achieve an agreement with the real estate agent. All this information describes your "position" or "requirements" in the negotiations. While you are doing your preparation, the other side is doing theirs. They have set fees and charges they plan to offer you so that they have the least amount of risk and the most profit. When you arrive at the real estate office, you discuss the costs and options and the agent gives up a few things to come to an agreement you can both live with. The contract is then completed, and the property is yours to use.

Be personally prepared before you go into a negotiation. Roger Fisher and Scott Brown tell us to address our emotions (Fisher, 1988). They suggest we prepare for negotiating by looking at the possible emotional responses that may occur in the negotiations. Run your position through a scenario and see what emotions may be triggered. Next, run what you believe the opposing position would be through a scenario and see what emotional triggers exist. When you have identified the triggers, determine ways that you can control your emotions. While you are in negotiations, do not say or do something that ends or significantly disrupts the discussion. Raising your voice or pounding your fist on the table makes for good television, but it can completely derail a negotiation. Keep your cool even when the other side says or does something to aggravate you. The best advice about negotiating is to "breathe." I know this advice may sound silly, but breathing with your diaphragm helps you maintain a calm disposition. You can also take a break if you get too wound up. Take a walk down to the restroom, get a drink of water from a fountain, or step outside for a breath of fresh air. After a few minutes, you return to the negotiations with renewed energy.

Negotiating

You should spend more than half of the time associated with a single negotiation on preparing. Once you have prepared to negotiate, show up ready to go, and this part is relatively short. Always open with the first offer, which is less than or more than the least acceptable outcome you identified earlier. If you were buying a car, you go under the price, but if you are bidding a job estimate, go above the estimate. This gives you the initiative. From that point, participate in the give and take of the negotiations. You should be listening twice as much as your speaking. This gives you the upper hand by allowing you to understand the other side's approach better. Be honest and forthright in your negotiations without trying to take advantage of the other side. If the negotiations do not seem to be moving in the direction you want, do not be afraid to end them. Too many people are unable to say "negotiations must end here." If the negotiations are not meeting your needs, it is perfectly okay to stop them and move on. If you allow yourself to be talked into a negotiation that does not give you what you need, you will be frustrated and mad at yourself. Not to mention the result will not be in your best interest.

While negotiating, do not say "NO" in response to something the other side raises. This alone can end the negotiations. Other forms of "NO" are: I cannot, we will not be able to, I do not see that happening. You are better off using the alternative "YES, IF." This allows you to swing the discussion back in your favor without compromising your goals. An example would be, "Yes, I can pay $2,000 for a car if you pay the taxes and license fees." In this case, you give a little more for the car while saving the money for taxes and licensing. If the other side does not agree, you are provided an opportunity to gain value. The best response is to give them something that they place a high value on, which does not cost you much. This puts you back on the offensive at little or no cost to you; however, never tell them it is at little or no cost to you.

How can you know if you have a successful negotiation? Roger Fisher and William Ury say that there are three criteria. "It should produce a wise agreement if the agreement is possible. It should be efficient. Moreover, it should improve or at least not damage the relationship between the parties" (Fisher, 1983). If the negotiations you are involved in do not accomplish all three criteria, you should walk away.

Regardless of how the negotiations turn out, it is important to part on good terms. If negotiations are successful, they usually end in a contract or some agreement. If the negotiations cannot meet both parties' needs, each of you should agree to stop negotiations and thank each other for the time and effort the negotiations took. Even if the other party says or does something that causes the negotiations to end, it is important that you take the high ground and thank the other party for the time and effort they put in the negotiations. Business relationships are more important in the long run than just the potential of the current negotiations, so take the effort to build and maintain them.

Concluding Negotiations

The final act of the negotiations is to get a commitment from both sides to the terms and conditions of a deal. If the other side does not confirm this, you most certainly should. Once you get agreement on the commitment, I recommend getting a handwritten note on the terms and conditions by both sides. The negotiations are over. When the paperwork arrives, review it carefully to ensure the terms and conditions have not changed from the negotiations.

After the negotiations are finished, hold a post-negotiation review to determine if you were properly prepared, knew your position well enough, knew the other side's position well enough, and remained in control while using proper emotions. Then determine whether you could have done things better or differently that would have improved the outcome of the negotiations. It is also polite to send a short note thanking the other side. Remember that relationships are often more important than the results of a single deal. All notes and records of the negotiations should be kept in a file that you can refer to for future negotiations with this same group. Before negotiating with this group again, you should review this file to remind yourself of what took place last time.

Summary

Negotiating is a skill that every SOH manager needs to master. Purchasing, leasing, and labor relations are just a few of the areas that a SOH manager must be able to negotiate. The good news is that all of us have some skill with negotiating, but we need to develop the skill professionally to be successful at work. It is easy to be nasty and shortsighted while negotiating, especially if the other side does it first. However, long term positive results come from treating others with dignity and respect. Furthermore, you should require others to treat you the same way. If you use these skills, you can be a successful negotiator and SOH manager.

CHAPTER 8 –INFLUENCING

Introduction

Safety and Occupational Health (SOH) Managers, in general, have limited power and authority. They are also provided with a small staff. If you take nothing else away from this chapter please know, "it is not the SOH manager's job to correct hazards." The person responsible for correcting hazards is the person responsible for the hazard. This could be the maintenance supervisor, facility manager, building engineer, human resources manager or business owner. Unfortunately, everyone in the company will look to the SOH manager to correct the hazards; however, the SOH manager and his or her staff cannot possibly get all these things done. This requires the SOH manager to influence others to help them. The others they try to convince are known as stakeholders.

This term is not usually found in the literature on SOH. It is not in any definition of terms; however, there is an explanation that comes very close. The Project Management Institute defines stakeholder as an "individual, group, or organization which may affect, be affected by, or perceive itself to be affected by a decision, activity, or outcome of a project, program, or portfolio" (PMI, 2012). From an SOH perspective, the definition of stakeholder should be "an individual, group, or organization which may affect, be affected by, or perceive itself to be affected by an unsafe or unhealthy working condition (hazard). This means that workers, supervisors, and managers in an organization are stakeholders. It also means that customers, the public, visitors, governments, and guests are also stakeholders. When a SOH manager needs help to get things done, he or she should reach out to stakeholders to help identify, control or eliminate the unsafe or unhealthful working conditions (hazards) within the organization.

It is at these times that the SOH manager must use influence, but what is influence? The Merriam-Webster Dictionary defines influence as the power to change or affect someone or something, the power to cause changes without directly forcing them to happen, or a person or thing that affects someone or something in an important way (Merriam, 2014).

Legacy Model of Influence

Perhaps the most famous teacher of influence is Dale Carnegie. In his book "How to Win Friends and Influence People," he provided the most eloquent descriptions of the application of influence I have ever read. I strongly encourage you to read this book. One of the things I like about Dale Carnegie's approach is that it relies on relationships. He stresses that it is about people understanding and encouraging each other. He does not support influence being used to take advantage of someone. In his book, Mr. Carnegie refers to influence as "How to Win People to Your Way of Thinking" and he has some core principles (Carnegie, 1936):

1. The only way to get the best of an argument is to avoid it.
2. Show respect for the other person's opinions.
3. If you are wrong, admit it quickly and emphatically.
4. Begin in a friendly way.
5. Get the other person saying yes, yes immediately.
6. Let the other persons do a great deal of the talking.
7. Let the other persons feel that the idea is his or hers.
8. Try to see things from the other person's point of view.
9. Be sympathetic to the other person's ideas and desires.
10. Appeal to the nobler motives.
11. Dramatize your thoughts.
12. Throw down a challenge.

As you can tell, these all flow from caring for a person you are trying to win to your way of thinking. If you start with genuine concern and interest in other people, they will be open and interested in what you must say. Through this method, you convince the other people that what you want to do is in their interest too. This gives them the desire to move in the direction you are advocating.

This is where influence begins. I have never seen anyone be successful influencing others unless they first took the time to build this relationship that is leveraged to gain the other person's support.

Alternative View of Influence

After you learn the standard definitions and applications of influence, you have a base from which to look for a method or methods that work for you. I suggest looking outside the box, which can result in a better alternative. In his book "Influence and Lead," Michael Nir describes an alternative view of influence. He has a new four quadrant grid that fits the topic very well. He uses the axes trust and agreement. Per him, stakeholders are divided into four groups: allies, opponents, accomplices, and adversaries. Michael Nir goes on to explain each of the quadrants. He says:

- Allies – advocates for you
- Opponents – openly and objectively against you
- Accomplices – outwardly accepting while giving you lip service
- Adversaries – do not accept and will not collaborate with you

More importantly, Michael Nir explains that accomplices and adversaries are "fence sitters" and can be convinced to support you or your adversaries. Two points I like about Michael Nir's take on this subject are:

> "In any interaction that you might have, approximately 5% of the stakeholders will be against, 5% of the stakeholders will be in favor, and the remaining 90% of the stakeholders will either be fence sitters or paying some amount of lip service" (Nir, 2013).

> "Do not forget that usually when trying to build a support coalition you will put too much emphasis on stakeholders who are against, which will lead to a failed effort" (Nir, 2013).

Successful stakeholder management is critical to building influence among stakeholders. You cannot do this unless you understand who the stakeholders are. This is another parallel to project management. In project management, the project manager gets to know each stakeholder. Through this effort, he or she understands what each one needs and how to provide it to them. It is also obvious which stakeholders have power and authority and which do not. This understanding is essential to meeting stakeholder expectations. It should be done the same way in SOH. By knowing the stakeholders involved, the SOH manager is in a better position to satisfy each one and gain influence at the same time.

SOH managers must also understand how to identify the stakeholders that support him or her and those that do not. Using this information in the way Mr. Nir describes allows the SOH manager to spend his or her time working to convince the right people, the ones that could support him with the goal of getting those stakeholders sitting on the fence to join in and become part of the solution. It is very easy for the stakeholders against an attempt to derail that effort. Focusing on preventing this does not seem to solve anything.

Some SOH managers make the mistake of assuming that since this involves controlling or eliminating hazards to prevent injuries and illnesses that everyone wants to join and help. That is not the case because there is a risk involved. Just because a risk exists does not mean it creates an injury today or next year. Part of seeking influence is to counter this unknown with a healthy dose of risk management. Providing an assessment of the frequency and severity of each hazard can help support efforts to correct them. This is just one way to package the hazards to help stakeholders understand them better and want to control or eliminate them.

Power of Influence

"People will change their behavior if they believe it will be worth it, and they believe they can do what is required" (MacMillan, 2013). This follows what has been said thus far. To get people to do things your way they need to know it is worth it. What does that mean? You have probably heard of WIIFM. That is an acronym for What's In It For Me? Me is, of course, the other person you are trying to influence. As a SOH manager, you need to explain to the people what's in it for them. This may take the form of a simple explanation or a short story. The people help if you can describe to them what they get for the effort. It is important that each individual you are trying to influence understands what they get. This is an area where SOH managers need to improve. It is not enough to get everyone to join in just because this effort is intended to prevent injuries and illnesses. Many people fall victim to the concept that "it will never happen to me" or anyone around them. SOH managers must counter this with a story that explains how bad things happen in any workplace. There are a lot of selfish people in the workplace, and some will not participate unless they get something out of it.

The second part of this is that the people you are influencing need to know they can do whatever it is you are asking. This not only means technically, but also whether they have the time. It is important to understand fully and can explain what it is you need people to do. Break it down into parts and pieces that can be done quickly with little effort.

Then it is easier to convince people that it can be done. Rather than highlight a list of 126 safety hazards, focus on the ones that need to be fixed right away and could be done with help in one or two weeks. Then move on to other hazards. Using this type of system does two things: 1) gives people work they can do in a period, and 2) allows people to complete something that encourages them to do more.

Influence creates synergy. Another way that influence produces power is by combining the effort of several workers, which adds up to more than they could each accomplish by themselves if you added each worker's effort together. Synergy is critical and helps the workers gain influence by working on teams. SOH managers can use this power to their advantage and get more done by a group than by using individuals.

Summary

SOH managers have limited power and authority and a small staff. In many workplaces, they are expected to correct or control every hazard. Again, if you take nothing else from this chapter please know, "it is not the SOH manager's job to correct hazards." Successful SOH managers ensure that every worker in an organization understands that they are responsible for correcting hazards. This includes everyone from the worker to the top official and all those in between. If the organization relies on the SOH manager to correct the hazards, it simply will not get done. Those that are not controlled or eliminated continue to place workers at risk. One important method to get others involved is to influence their behavior. If everyone joins in, all hazards can be controlled or eliminated and accidents prevented.

CHAPTER 9 - DECISION TOOLS

Introduction

As a Safety and Occupational Health (SOH) Manager, you should have a college degree or at least completed some college classes. Along with that, you should have completed a variety of technical training from the basics of safety to more advanced training in accident investigation and human factors engineering or ergonomics. Along the way to completing that education and training, you should have learned a lot about making decisions and presenting data to support conclusions. I do not want to tell you about those things that I think you already know. Just to level the playing field, the things I believe you have or should have already learned include:

- The difference between qualitative and quantitative data.
- How to determine the Mean, Median, Mode, Range, and Standard Deviation.
- How to use run charts, flow charts, histograms, Pareto charts, and tree diagrams; force field analyses, cause and effect diagrams, brainstorming, Matrix Diagrams, and Check Sheets.

Instead, what I would like to do is address those things I do not think you know, but that I believe can help you do a better job.

SOH managers usually are not able to make decisions for hazard elimination and control. For those decisions, we often go to someone in management who has the authority and financial resources to decide such things. To do this effectively, the SOH manager needs to be able to show cause and effect between the controlling or eliminating the hazard and reduced risk or cost in the future. I am referring to return on the investment of money spent on the danger. This should be done regarding options that a manager or executive can choose from. Furthermore, the SOH manager must be able to determine the cost of options accurately concerning current dollars, future dollars, lifetime costs, and internal rates of return. Understanding the time value of money is important if you want to get approval to spend that money to control or eliminate hazards.

Net Present Value

Many projects to control or eliminate hazards are expensive. They are identified as Capital Projects, which means that the company spends a considerable amount of money today with the expectation of future returns through additional income or reduced costs (Finance, 2013). Net Present Value (NPV) is defined as "the monetary value today that an investment project earns after yielding the desired rate of return for each period during the life of the investment" (Finance, 2013). The formula that I use for NPV is:

NPV (Net Present Value) = $FV/(1+r)^n$

For this formula, FV= Future Value, r= interest rate, and n= number of time periods. The higher the NPV, the better. Business managers want the money that is being used for correcting safety hazards spent wisely. When recommending measures to control or eliminate hazards, it is important to identify the cost of that money as Net Present Value to allow the financial staff to compare future returns on the money. The interest rate comes from the financial staff of your organization.

Internal Rate of Return

The International Facility Management Association defines Internal Rate of Return as the interest rate at which lifetime dollar savings equal lifetime dollar costs if the time value of money is considered. This rate is then compared to the minimum acceptable corporate rate of return to determine if the investment is desirable (Finance, 2013). You may also see Internal Rate of Return conveyed with the acronym IRR. The formula that I use for IRR is:

IRR (Internal Rate of Return) = Solve the Net Present Value equation for "r".

The higher the IRR percentage, the better. This is because most organizations borrow money to get work like fixing safety hazards done. Borrowing money always comes with the cost of a certain rate. Your IRR must be higher than that rate, or the organization loses money. Even if the organization does not borrow money it often invests money. If the money to fix a safety issue is taken from the money the organization plans to invest, then the IRR must be higher than the rate of investing. This comparison of the value of rates helps the organization get the most use of its money. Furthermore, a business manager uses this rate to compare different projects. The one with the highest IRR is given the money to proceed while the others usually must wait.

Interrelationship Diagram

An interrelationship diagram illustrates cause and effect relationships among many issues. It is used when there are several causes of variation, and you are unsure which cause has the most effect on the others.

This is a tool for a team to use. Once the main causes of a problem or variation are identified, write them on sticky notes and place them randomly on a large piece of paper. Draw arrows between two items that are related from the cause to the effect. The primary cause is the one that has the most arrows coming from it.

What is the leading cause of sick building syndrome? See Figure 1 for an example.

There are several arrows that come from different boxes; therefore, there appear to be several causes of Sick Building Syndrome. It should also be noted that a number of arrows go to improper ventilation, indicating that it is the root cause of several of the symptoms. Use this when you need to identify the connections between causes to allow you to understand better and explain how the measures work to correct the hazards.

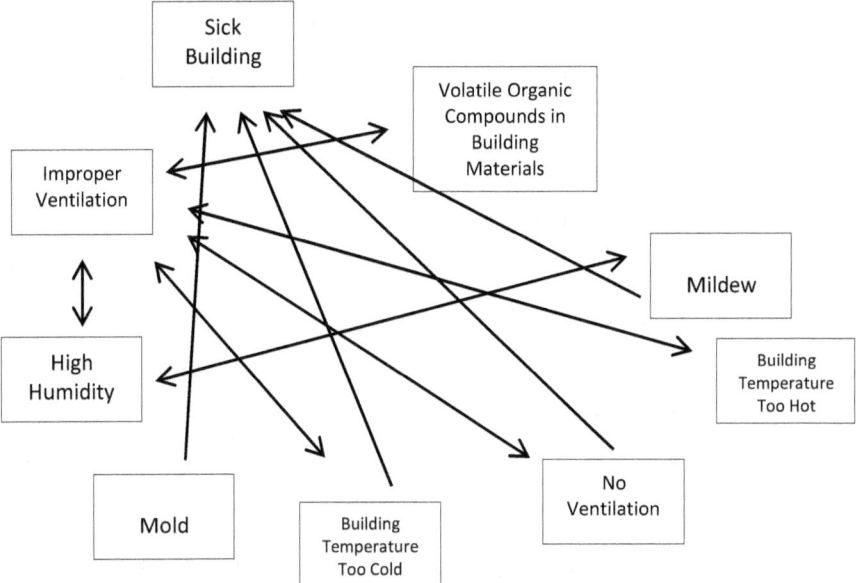

Figure 1 - Interrelationship Diagram

Force Field Analysis

The force field analysis is necessary because it helps you recognize the forces that can help you do a task as well as the forces that prevent you from doing the task. It is important to know these forces as well as ensure management knows them too. You should use the driving forces to help complete the task while controlling or eliminating restraining forces. Management may be needed to help control or eliminate forces against your task. It is usually considered most useful to enhance the driving forces than to remove the restraining forces. An example is in Figure 2 below.

Goal: Get Workers to Wear Hard Hat

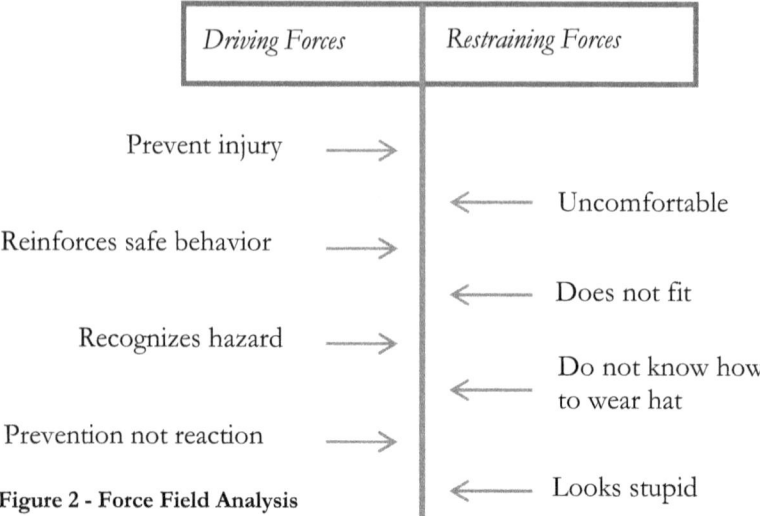

Figure 2 - Force Field Analysis

Decision Matrix

The purpose of a decision matrix is to help SOH managers make better decisions by taking some of the subjectivity out of the process. There i a variety of decision matrices that can be used. Decision matrices require the identification of possible alternative decisions that can be made as well as the criteria that all decisions must meet to be considered. I like the decision matrix developed by the Army Management Staff College known as the Army Management Staff College – Mission Execution and Decision System or AMED. An example is in Table 1. To make this information more applicable, it is important to determine the criteria under which the issues could be resolved. Measure the impact of each of these criteria on a scale of +5 to –5 with +5 being the best, 0 being no change, and -5 being the worst. You then identify the alternative with the greatest potential to succeed. The last column is where you put the restraining forces. If we go back to our example of sick building syndrome, it will look something like this:

Alternative	Weighted Criteria				
	Increase morale	Reduced Hazards	Improve HVAC system	Total	Restraining Forces
Do nothing.	0	0	0	0	0
Adjust ventilation so that it provides enough air supply based on engineering analysis	0	+3	+3	+6	-1
Increase building temperature.	+2	+4	+3	+9	-2

Table 1 – Example AMED Matrix

In this example increasing building temperature is the best alternative based on the sum of the criteria minus the restraining forces.

SWOT Analysis

A SWOT analysis is commonly used in business and should be used more often by SOH Managers. The acronym SWOT stands for Strengths, Weaknesses, Opportunities, and Threats. SWOT analysis is a method of analysis using four elements. Figure 3 is an example of a SWOT diagram used in the analysis.

	Helpful	Not Helpful
Internal to Organization	Strengths	Weaknesses
External to Organization	Opportunities	Threats

Figure 3 – Typical SWOT Diagram

A SWOT analysis is one of the most widely used strategic analysis tools (Trevor, 2014). I used them frequently throughout my working career. I found the analysis to be very effective and useful. In some cases, people have trouble with boiling things down to strengths, weaknesses, opportunities, and threats. To help better understand these terms, Rayne Hall (2015) refers to the strengths as internal positives and weaknesses as internal negatives. This is an excellent way to describe them. She also refers to the opportunities as external positives and the threats as external negatives (Hall, 2015). I also like to use these descriptions.

I like to use the following descriptions when using the SWOT analysis:

- Strengths: my knowledge, skills, and abilities. Things I do well.
- Weaknesses: areas in which my knowledge, skills, and abilities are lacking. Things I don't do well.
- Opportunities: people, places, and things external to me that will help me reach my goal.
- Threats: people, places, and things external to me that will prevent me from reaching my goal.

To do a SWOT analysis, you develop a list of each of the strengths, weaknesses, opportunities, and threats and fill in the squares of the matrix. For example, let's look at presenting a safety class in Figure 4.

If the SWOT diagram has more entries in the strengths and weaknesses blocks, you should pay more attention to the external factors of opportunities and threats. If the SWOT diagram has more entries in the opportunities and threats, you should pay more attention to the internal factors of strengths and weaknesses.

If the SWOT diagram has more entries in the strengths and opportunities blocks, you may not have been fully honest about your weaknesses and the threats around you. You should put more effort into finding threats and weaknesses. If the SWOT diagram has more entries in the weaknesses and threats blocks, you may not have been honest about your strengths or the opportunities around you. You should put more effort into finding strengths and opportunities.

The goal is for the SWOT to have about the same number of entries in each of the four blocks. Ryan Hall (2015) provides a better explanation in her book for validating the balance of the SWOT diagram.

	Helpful	Not Helpful
Internal to Organization	- I know the material - I am good at presenting. - I have a good room to use for presentation - I have good slides to show - I have good handouts	- I don't know the employees. - There supervisor might not want to give me the time. - The employees may not want to attend presentation.
External to Organization	- I have management's support. - I have been properly trained. - I belong to a society with members that can help.	- The union will want to have a say in what is presented. - Employees my call in sick that day to avoid presentation. - Employees may ban together to ignore training.

Figure 4 – Example SWOT diagram for Training Presentation

After completing a SWOT diagram, the next step would be to expound on each entry. For example, in Figure 4 in the Strengths block, you listed "I know the material." You could add a subentry such as, "Include a question and answer period after the presentation to make sure those attending understand." We can also try a negative entry from Figure 4. In the Threats block, you said, "The union will want to have a say in what is presented." You could add a subentry such as, "I will contact Management Employee Relations to work the union piece." You add subentries to each of the entries in the four blocks. Once all the subentries are made, go back and identify the ones you would like to do and highlight them in yellow. Now go back to each subentry you highlighted in yellow and write an action statement that you will take to do the subentry. This will give you an action plan to take advantage of the Strengths and Opportunities while avoiding negative impacts from your Weaknesses and Threats.

Some people find it easier to write lists than use the SWOT diagram. The point is not the format, but that you look at each of the four elements and end up with a list of actions to reach your goal.

Summary

SOH Managers must know a lot about making decisions and presenting data to support conclusions. I have provided examples I think you should be familiar with here. The time spent on learning these methods is time well spent.

CHAPTER 10 – BODY LANGUAGE

Introduction

In addition to spoken words, we often use other ways of speaking to communicate. Unfortunately, people can understand or misunderstand those other ways of communication. To be successful in the other ways, it is important for us to ensure our messages are understood. The other ways of speaking are known as nonverbal communications. Body language is one form of non-verbal communications we use to communicate with other people. You may be shocked to know that many of the ways we communicate through body language are done without our even knowing it. This means that we can deny or contradict our verbal language with our body language. This can be dangerous but never more so than when speaking to direct reports, bosses, or customers. In these professional situations, it is important to make sure your body language agrees with your verbal language. In professional situations, it is also important to know what the body language of a direct report, boss, or customer is saying to you and whether it agrees with their verbal language.

Where to Begin?

Tonya Reiman is a body language expert, author, and frequent television guest. Reiman says, "Research has found that as much as 93 percent of our interpersonal communication is nonverbal" (Reiman, 2007). Safety and Occupational Health (SOH) managers communicate with a variety of people using interpersonal communications. In fact, you could say our very existence depends on our ability to communicate with other people up and down the corporate ladder and with customers and vendors. If 93% of our communications are non-verbal, we need to learn them correctly. Reiman goes on to explain, "Our conscious brains might be focused on decoding the spoken words in the conversation, but the subconscious does the really heavy lifting, "reading" the body's many languages for nonverbal cues that tell us about the other person's real intentions" (Reiman, 6, 2007). That teaches us that we not only need to learn how to communicate using body language, but we need to know how to read the body language of the other person communicating with us. In this way, we can better understand the full message.

David Lambert, another body language expert, states in his book Body Language "that it seems to have three broad uses: as a conscious replacement for speech; to reinforce speech; and as a mirror or betrayer of mood" (Lambert, 1, 2007). This is like what Tonya Reiman says in her books. Of these three uses, we want to avoid our body language betraying our mood as well as recognizing when another person's body language betrays their mood.

What does this betrayal look like? You may have heard a person say, "Yes, I agree" but they shake their head from left to right indicating "No." What does this mean? Well, the person could be:

- Lying
- Are not sure if the answer is yes
- Might not want to hurt the other person's feelings (another form of lying)
- May not have made their mind up

In any of these cases, it is essential for you to know what this person means or believes always, and that is "No." It may not be necessary for you know what this person means in all cases. If this is an employee that directly reports to you and is answering questions about a workplace theft, you want to know what he or she means to say.

Two lessons that SOH managers need to know are that people communicate by other methods than verbal communications and that what a person says and how his body is acting should be congruent or similar. If the messages are different, it does not mean the other person is lying, but it means he might be. However, if the verbal and nonverbal is similar, it is likely the other person is telling you something truthful or at least believes what they are saying is true.

A Few Basics

In a short chapter, it is impossible to cover the topic of body language completely; however, I would like to provide you with the high points. If you would like to learn more, I highly recommend Tonya Reiman's book *The Power of Body Language* and David Lambert's book *Body Language 101*. David Lambert includes a number of pictures in his book, which makes it easier to understand.

The first area you should become familiar with is facial expressions. Facial expressions should agree with the words the person uses. "There are at least six facial expressions found throughout the world, which would suggest that they are inborn rather than learned. They are happiness, sadness, surprise, fear, anger and disgust" (Lambert, 2007). It is important to recognize each of these facial expressions so that you can compare them with the words used by an individual showing this expression.

"Apart from the face, the most visually expressive features are the hands" per David Lambert (2007). Power grips, symbolic thrusts, and open hand gestures are just a few of the most important uses of hands. The use of the pointed finger can undermine your point by intimidating the other person, unless that is what you intend to do. Placing your palms up in a gesture means you're open while pushing your arms out with your palms towards the other person distances you from the other person.

People have three areas of distance, but personal space is the one we need to consider for professional communications. Have you seen a person back away as a person steps up to them and begins to speak? That usually means the second person violated the personal space of the first person. "Professional Use of Space and Crowd Behavior Space is relative" (Reiman, 2008). We each have different amounts of personal space we protect. The furthest point from us is usually for public meetings and discussions with people we do not know. The next space closer in is for social events, perhaps a party when two people casually speak. The next in is for an intimate discussion like that of a couple or married couple. If you violate personal space, the other person becomes uneasy and often backs away. It is important to respect personal space and thereby keep the focus on the information not the violation of the personal space. A person may violate your personal space on purpose to distract you from your point. It is important to understand what the other person is doing when this occurs.

Mirroring the other person's body language can help you quickly gain a common understanding. The overall effect of mirroring is to create the impression of being on the same wavelength as another (Reiman, 2008). If the other person is sitting, then you sit in a very similar way. If the other person leans forward, then you lean forward. Remember to do this with the intent of improving your communication with this other person and not to take advantage of them. Also, take note when a person does this to you. They may truly be on the same wavelength as you or they may be mirroring you to gain an advantage.

First Impressions

Tonya Reiman wrote an excellent article titled Those First Few Seconds. In her piece, she states, "Everyone knows you are supposed to make a good first impression. Unfortunately, no one ever gave us an instruction manual for making a knock-em-dead first impression" (Reiman, 2008). She is right. We should have received training on something this important. Before we jump into how to do it right, it is important to know why we give first impressions so much weight.

The Psychology Today magazine website tells us, "We are built to size each other up quickly. Even if we are later presented with lots of evidence to the contrary, we are attached to our initial impressions of people—which is why you should be aware of the impression you make on others. Luckily there are simple guidelines for wowing new acquaintances" (Psychology, 2013). The first area we as SOH managers should focus on is making sure we say and do the right thing when we first meet someone. This means our language and nonverbal cues should be congruent.

First, you should always be genuine in your first impression. Do not pretend or lie, or your body will most likely give you away. Begin with being properly groomed and dressed. Say what you mean and do not ad lib or say something you do not agree with. Look the other person in the eye, and if shaking hands is appropriate for the culture, shake and mean it.

Consider what the other person looks like. Not being properly groomed or dressed well will negatively affect the way you think of the other person. Listen to what the other person says and see if his or her body language is congruent with the words they use.

Cultural Differences

What I have included here is for body language in Western societies. Body language often changes with the culture of the users. Here we shake our head from left to right to mean disagreement, but in other countries, this same gesture means yes. SOH managers are now more world focused than ever before. As you travel, you must first identify the proper gestures used in the country you are going to and use them appropriately. When a person from a different culture comes to your location to speak to you, it is also important to use the proper gestures and understand their meaning. This prevents misunderstandings that could ruin a business deal or a business relationship. Neither of which is good for a SOH manager.

Summary

As SOH managers, we use verbal and nonverbal language to communicate. It is, therefore, important that SOH managers understand how important it is for verbal and nonverbal language to be congruent or similar as well as culturally correct. This allows SOH managers to be successful in the ways they speak to other people with messages that are understood. You can become familiar with body language through the references I have included in the bibliography. I hope you take the time now to get familiar with body language. I think you will be a better SOH manager for the effort.

CHAPTER 11 – EMOTIONAL INTELLIGENCE

Introduction

Do you know a Safety and Occupational Health (SOH) manager who has lost their temper and lashed out at an employee verbally? Do you know a SOH manager who's responded emotionally to a customer, making an unreasonable demand? Unfortunately, many of us react poorly or overreact to situations before we realize it. I know a SOH manager who was receiving an update from an employee along with the employee's supervisor. The employee reported a slippage in the project timeline. The SOH manager blew up and began screaming at the employee that this or any other slippage was unacceptable. Without warning, the SOH manager picked up a binder from his desk and threw it, just missing the worker's head. The supervisor intervened and suggested that the meeting be reconvened the next day. The SOH manager, realizing his mistake, agreed to take up the meeting the next day. The employee lost control of the project schedule, but the SOH manager lost control of himself. The relationship between the SOH manager and the employee was nearly destroyed.

What is Emotional Intelligence?

The Glossary of the Project Management Body of Knowledge (PMBoK) Guide (2013) defines emotional intelligence or EI as "the capability to identify, assess, and manage the personal emotions of oneself and other people, as well as collective emotions of groups of people." There are other definitions, and Anthony Moberg (2011) stated in his book Emotional Intelligence: An Introduction, "In general terms, Emotional Intelligence (EI) refers to the ability to perceive, control, and evaluate emotions – your own and those of other people." To take that one step further, Daniel Goleman states "…emotional intelligence determines our potential for learning the fundamentals of self-mastery and the like; our emotional competence shows how much of that potential we have mastered" (Goleman, 2005). There are three recognized EI models, but only one will be addressed in this book. The Trait EI Model is focused on regulating emotions and personal growth and is most appropriate for helping SOH managers improve how they deal with emotional situations" (Moberg, 2011).

SOH managers, just like anyone else, have different responses to emotions, and sometimes the reaction is out of place or detrimental to the situation. "It is only through exercising emotional intelligence that we can manage these responses more rationally..." (Morley, 2011).

Know Thyself

EI focuses on the brain. The brain is the organ that, at least in part, controls emotions in humans. There are two parts to the human brain, the Emotional Brain (Amygdala) or EB and the Thinking Brain (neocortex) or TB. The EB processes the same information as TB. The EB generates emotions subconsciously, processing the information more quickly than the TB can respond with logic (Diamond, 2011). That is why the SOH manager in the earlier example starts to yell and throws things then realizes his actions were wrong and agrees to end the meeting. Anthony Moberg (2011) calls this "emotional hijacking...when your emotional brain takes control, subverting your rational thinking response." Emotional hijacking can happen to anyone, but it has a greater chance of happening to someone who does not have control of their emotions. Contributing factors could be stress, strain, or fatigue. Each of us needs to identify when we are emotionally hijacked "when we start to feel drained, frustrated, irritated, angry, sad, fearful, or any emotion that has no place in a professional work environment" (Moberg, 2011). It happens differently to each of us, and we must learn how it happens. I have found that recognizing an emotional hijacking as it first begins to happen is the key to responding properly. Next, you need to know what sets you off. A person at work, a situation, a family member, or time of day can be what sets you off. I have a high maintenance employee who does the craziest things that create emergencies for me that set me off. Many SOH managers probably have the same situation. Lou Diamond (2011) says that you must also be socially aware, which involves "knowing how you react to social situations." The best way to do this is to analyze how you respond by being in social situations. You must also speak to those close to you and find out what they know of your reactions, triggers, and over the top responses. I do not have to tell you that learning these things about yourself may not be easy, but knowing them is the only way you can take charge of your EI. I recommend taking a piece of paper and making four columns on it. In the left column, identify the emotions you have recognized along with the conditions or the people who make you respond.

Improve Yourself

Daniel Goleman quoted Aristotle with what I think is the best summary of this subject: "Anyone can become Angry-That is easy. But to be angry with the right person, to the right degree, at the right time, for the right purpose, and in the right way-that is not easy" (Aristotle, The Nichomachean Ethics). I sincerely believe that EI can help SOH managers have the emotion they need with the right person, at the right level, at the right time, for the best idea, and in the best way. After we take stock of who we are and how we react to situations with other people, we can find ways to respond better.

You have probably heard of holding your breath and counting to ten before responding to things that make you mad or angry. I have found that works very well. Another technique is not responding until 24 hours after being confronted with something that makes you mad or angry. I have also used that method with great success. Lou Diamond (2011) "recommends taking a few deep breaths then facing your anger, anxiety, frustration, or whatever emotion you are feeling to practice your emotion response ability." You probably already realize that this can be referred to as self-realization. Whatever you call it, each of us must know more about ourselves and how we respond to situations that occur around us. I recommend going back to the list you developed in the emotions, situation, or people who make you react. In the second column, identify a way that you can recognize the emotion as it is starting to develop. In the third column, add triggers that can exacerbate your response. In the last column, identify what method you use to control your response. Practice can increase Emotional Intelligence.

Summary

"SOH Manager- Know Thyself" is a mantra that each of us should come to live by. Many of us react poorly or overreact to situations before we realize it. These reactions can damage relationships we have with customers, direct reports, and our supervisor. The stress and strain we are under nowadays can make these conditions worse. We can first take stock of who we are and how we respond to people and situations. We should learn what triggers our responses. By knowing these things about us, we are better able to identify steps we can take to regain control of our emotions and appropriately respond in all conditions. SOH managers today must be familiar with EI. I hope you found this subject interesting.

CHAPTER 12 – EMPLOYEES AND EMOTIONAL INTELLIGENCE

Introduction

Have you seen an employee lose their temper and lash out at another employee? Have you seen an employee respond inappropriately to the customer? Unfortunately, incidents like these occur in workplaces every day. I know of an employee who assaulted a custodian in the office. As the custodian stepped over a telephone cord, the employee jerked it up between her legs and laughed heartily. The employee, who claimed it was just a joke, was found guilty of assault and punished with several days off work without pay. To make matters worse, when he returned to work the female secretary was afraid to be alone with him. The relationship between the employee and his peers was destroyed. Employees need a deeper understanding of what emotional intelligence is to understand why they act the way they do and enable them to improve how they respond to others around them.

Why Emotional Intelligence?

In a previous chapter, I familiarized you, the Safety and Occupational Health Manager, with Emotional Intelligence. I repeat that information here to say that in the workplace we want employees to respond with emotions that are appropriate for the circumstances at that time and to be able to recognize what other people are feeling. What you get for this effort is a worker who is better able to regulate their emotions, whichreduces issues within the office and helps your employees be more satisfied. Cherniss (2000) refers to this as learning "Empathy" that contributes to the worker being successful. It has been found that if an employee does not have empathy or does not respond with empathy at the appropriate time, workplace relationships can be damaged, resulting in less success at work. The Trait EI Model that you would use for yourself is also the best model to use for employees (Moberg, 2011). The bottom line is that employees must respond with the proper emotion at the proper intensity; however, they cannot do that unless they understand how they react and why it is or is not acceptable for the situation and then be taught how to respond appropriately. This requires you train and coach, so employees learn how to respond.

Helping Employees to Know Themselves

The employee I referred to earlier committed what he thought was a joke that ended up with his being found guilty of assault. He acted in this situation with his emotional brain rather than his thinking brain. So he did what he thought was a joke, without regard to what the cleaning lady might think about it. This is what Anthony Moberg describes emotional hijacking as, "When your emotional brain takes control, subverting your rational thought response" (Moberg, 2011). Emotional hijacking can happen to anyone, but it has a greater chance of happening to someone who does not have control of their emotions. The feeling of stress, strain, or fatigue can make the hijacking feel worse.

Employees need to learn when they are emotionally hijacked "when we start to feel drained, frustrated, irritated, angry, sad, fearful, or any emotion that has no place in a professional work environment" (Moberg, 2011). It happens differently to each of us, and employees must learn how it happens to them. I have found that helping employees recognize an emotional hijacking as it first begins is essential to responding appropriately.

Next, you need to help employees find out what sets them off. A person at work, a situation, a family member, or time of day can be what triggers them. Like many of you, I have had a high maintenance employee who does the craziest things that create emergencies for me that set me off. I must learn how to respond to this employee appropriately and prevent being hijacked before replying to him. That is what we need to teach the employee.

Lou Diamond says that we must also be socially aware. He refers to "knowing how you react to social situations" (Diamond, 2011). The best way to do this is to teach employees how to analyze how they respond by putting them in social situations. They must also speak to those close to them and find out what they know about the employee's reactions, triggers, and over the top responses. I do not have to tell you that teaching these things may not be easy, but knowing them is the only way an employee can take charge of their EI.

Next Steps

The first thing I recommend is that you provide a short familiarization course to your employees on Emotional Intelligence. There are a number of providers of this type of training. I suggest you make all employees attend and get their feedback on the experience. After this happens, you should conduct coaching sessions with the employees. At the first meeting, I suggest each employee take a piece of paper and make five columns on it. In the left column, have them identify the emotions they recognize in themselves and the conditions or the people who may set them off or cause them to respond inappropriately. Follow this session with additional coaching sessions.

Lou Diamond recommends taking a few deep breaths then facing their anger, anxiety, frustration, or whatever emotion you are feeling to practice your emotion response ability" (Diamond, 2011). You can refer to this as self-control. Whatever you call it, this is another good meeting topic for employees. I also suggest going back to the list the employees developed of the emotions, situations, or people who set them off. In the second column, have them identify a way that they can recognize the emotions as they are starting to develop. In the third column, have employees add triggers that can exacerbate their responses. In the last column, have them identify the method(s) they can use to control their reactions. Have them practice daily.

Summary

I sincerely believe that EI can help employees demonstrate the emotion they need to the right person, at the right level, at the right time, and in the best way. If we can help the employees take stock of whom they are and how they respond to situations with other people, they can find ways to respond better.

"Employee - Know Thyself" is a mantra that each employee should come to live by. Many employees react poorly or overreact to situations before they realize it. These reactions can damage relationships they have with customers, fellow employees, and you, their supervisor. The stress and strain employees are under nowadays can make these conditions worse. They first need to find out how they respond to people and situations. They need to learn what triggers their responses. By knowing these things about themselves, they are better able to identify steps they can take to control their emotions and appropriately respond under any condition. Employees must be familiar with EI and use it to help better themselves. As their supervisor, you can help them do it. I hope you found this subject interesting. As stated in the previous chapter, if you would like additional information, I recommend you read Emotional Intelligence by Daniel Goleman.

CHAPTER 13 – PERSONAL INTELLIGENCE

Introduction

In today's fast-paced world, Safety and Occupational Health (SOH) managers strive to hire winners that can succeed. These people can also push others in the organization to succeed. Unfortunately, it is hard to tell among a pool of applicants which is the winner. One theory that can help is Personal Intelligence (PI). Catie Hill (2014) notes that "From the summer of 2007 until early 2014, when John Mayer's book was published, a new psychological theory was born: the theory of personal intelligence, wherein John Mayer tried to explain how much we understand ourselves helps us interact with others. The theory was a follow on to emotional intelligence. The theory is defined as "personal intelligence is *an intelligence that involves reasoning about personality and personality-related information.*" Each of us has a personality, and personal intelligence allows us to reason both to ourselves and other people" (Personal Intelligence, 2014). As a hiring manager, there is a way to measure PI to inform the hiring process.

Why PI?

Even though PI is a theory, it has been proven in academic research. John Mayer along with A.T. Panter and David Caruso (2012) developed and conducted tests of individuals that identified PI as a separate intelligence that can be measured. There are a number of benefits that an employer can derive from employees with PI. "One key to personal intelligence is the ability to distinguish our perception of another person from whom the person is - or, in this case, was" (Mayer, 2014). At work, we often act differently than normal. This is done to create a defense against some fear that we have of not being smart enough or handsome enough. Even though this provides a level of protection for us, it makes it hard for others to develop a relationship with us. Being able to see beyond this to the real person gives an employee with high PI the ability to meet the needs of others and create relationships that facilitate good working and highly productive teams. John Mayer (2014) goes on to say that "to succeed, our personality must guide our actions in each of these areas- and as we act; we leave behind traces of who we are." Sharing these traces allows others to know more about us and interact with us better. This can be especially important in today's environment where work is often done by teams rather than individuals. Working as the member of a team requires us to know more about each other to improve our interactions.

The traits that people with high PI have also are effective when conducting SOH work that is contentious or requires good personal interaction. By understanding the personality of the other person, the SOH specialist can often respond better. Specific situations that a SOH specialist can benefit from high PI include conducting:

- SOH training
- Inspections
- Job Hazard Analyses
- Accident Investigations
- Safety Awareness Activities

SOH requires professionals with excellent people and communication skills. Having a high PI helps in both areas. One area of great importance is getting SOH work done through others. A major part of this is to gain the trust and cooperation of others, so they want to help prevent accidents by eliminating or controlling hazards.

Over the years, I have worked with SOH specialists and managers that did not have the skills of high PI. They intimidated people, said the wrong things, started arguments, created conflict instead of relationships. Some of these people meant to do this, but most did not. They were not able to understand their personality or the personality of those they worked with. Thus, there was no cooperation, and teamwork was marginally successful at best.

Unfortunately, this is not all good news. As you might expect, there can be issues for people with high levels of PI. "Individuals who are high in PI also may suffer from certain vulnerabilities. Persons with such skills may be more open to criticism than others—openness to criticism is part of developing self-knowledge—and as a consequence, he or she might face an increased risk of depression or dysphoria at some point in his or her life. Given his or her skills at self-guidance, however, he or she is likely to puzzle through such issues (making use of psychotherapy if needed)" (Mayer, 2014).

How to Use PI

"Per the theory, people high in PI can carry out problem-solving about personality better than others. Anyone who consistently exhibits this ability possesses personal intelligence (Mayer, 2014). This intelligence should allow the SOH specialist to understand better and react to other people's personality. This can help in creating the relationships necessary to be successful at work. The best way to use PI is to identify it as a criterion for employee selection. "A similar new theory - that of "personal intelligence" - may yet add more weight to the argument that a holistic approach to candidate recruitment and selection is needed" (Consultancy, 2014). This could be one of the many selection criteria used to screen potential job applicants. How an applicant measures on each criterion leads the manager to the point where the highest qualified candidate can be identified and then selected.

Another criterion I recommend is Emotional Intelligence or EI. Even though they appear to be similar, PI is different from EI.

The Personal Intelligence website has a quiz that can determine the amount of PI that a person has. There are additional resources on that same page to prepare managers for determining PI. The Human Resources Office in most organizations may not be aware of this new approach and method to better hiring. You can refer them to this website. The time spent early in the hiring process can eliminate hiring employees that do not have the qualities needed to fulfill the job duties.

Summary

Picking winners is the key to managing a successful SOH program and office. The impact that selecting employees with high PI can have on relationships with others is paramount. Hiring people with high PI can create work atmospheres in which SOH specialists have productive interactions with individuals in the workplace. Through these relationships, workers, supervisors, and managers take responsibility for the safety program and work to eliminate or control hazards. The website "Personal Intelligence" can be found at http://personalintelligence.info/. This is a great resource to learn more about PI as well as identify ways that you as a SOH manager can use it.

CHAPTER 14 – NEURO-LINGUISTIC PROGRAMMING

Introduction

Safety and Occupational Health (SOH) managers are people first and professionals second. As people, they have a disposition or personality that they demonstrate to other people. SOH managers are usually outgoing people, interested and concerned about others, wanting to make a difference in the world. Unfortunately, they also have some warts and foibles that afflict us all. Many times, they suffer from stress, experience depression and anxiety, and have addictions. Others put themselves down, not believing they can do the job or are good enough to do the job.

We all have self-talk. These are the discussions we have in our mind about who we are and what we do. You might be at a meeting with a new co-worker you met yesterday and mispronounced their name. Afterward, your internal self-talk might be about how stupid you are for not remembering that worker's name or perhaps even worse. In many cases, we are our worst enemy.

What we say to ourselves as well as the people around us is important. Words do matter. If a young person is told often enough that they are stupid, they begin to believe it and can live the rest of their lives with that self-image. There is a better way. Neuro-Linguistic Programming can help SOH managers learn the tools and techniques to identify their goals and objectives, find out how to describe them in favorable terms, develop a plan to reach the goals and objectives, and have the successful career they want.

The world-renowned performance coach Tony Robbins is just one of many who teach Neuro-Linguistics Programming. In this chapter, I am not going to teach you how to become rich or walk on beds of fire as he does. What I would like to do is give you some tools to help you achieve success as a SOH manager.

What is Neuro Linguistics Programming?

Neuro Linguistics Programming or NLP is a self-improvement program developed in the 1970s. George Lynch (2014) explains in his book that "Businesses, as well as individuals, have picked up on how useful NLP is in creating confidence, changing harmful thought patterns and controlling emotional states that were previously damaging to teamwork, job performance, and setting, as well as reaching life goals." That is the context I address using NLP for SOH managers. It is not important to dig deep into the definition for NLP because it goes off into the subconscious and conscious minds. My definition is that NLP is a method of reprogramming your mind to react to language in a way that is productive for you as an individual or as a professional. It begins with your beliefs and values. Your parents and other adults around you provided you with certain beliefs and values at an early age. You can change them, but it takes work. First, you have got to recognize what your beliefs and values are. Many of those values are good, but you may have some that are holding you back or getting in your way at work. A couple of examples I have noticed at work include:

- A young man was taught that working with your hands was the best way to earn a living. Unfortunately, he wanted to go to college and join the safety profession. His family did not support him and did not recognize his work, education, and certifications as necessary and the young man felt like he had failed because his family did not respect him. The young man stopped listening to his family and started to change the way he thought about himself. He knew his decision to go to college was the right decision, so he told himself that frequently. He hung his diploma on the wall of his office and every time he looked at it, he told himself he was proud. Several months later, his disposition at work improved and those working around him noticed a better attitude and personality.

- The second young man was raised in the belief that all women should be homemakers and care for the home and family. At work, he treated women differently than men and often told those he worked with closely of his opinion. The young man was counseled by his supervisor that his personal beliefs were his own, but that not everyone shared them, and he could not treat the women he worked with any different than the men just because the women chose to work. The young employee could not or would not change his behavior, and he was disciplined for his actions and statements that disrupted the workforce. The entire organization was affected.

Through knowing what your beliefs and values are, you can decide which to keep in your personal life, your professional life or not to keep at all. It is fine to respect your parent's beliefs, but still have a different belief of your own. It is also okay to have a personal belief and then a little different version of that same belief professionally. In the first example, a young man identified that his parent's belief was causing him to be unhappy that resulted in behavior at work that hurt the opinion others had of him. Once he realized he could change the way he felt and acted to reprogram his thinking, he could come to terms with his parents' beliefs and yet have his convictions. This improved his demeanor, and it was noticed in work. In the second example, a young man could not or would not see that his beliefs were causing consternation in the workplace. Others did not share his beliefs, and his efforts to force his beliefs on others caused a rift in not only his office but in the larger organization. This resulted in a consistent pattern of poor performance followed by discipline for this young man. Had this young man just held his personal beliefs while allowing others to have theirs, he could have been successful at work.

To summarize, the definition of NLP is that it is a method of reprogramming our mind to react to language in a way that is productive for you as an individual or as a professional. That language is in our mind as well with others we speak to, speak with, and hear. We all talk to ourselves. Sometimes we do not say very nice things, and we demean ourselves or hold ourselves to too high of a standard.

Clarence Rivers (2013) says, "You are your own worst enemy." He says that we sabotage what we do because we are afraid to fail. When we fail in trying, we can build a fear of failure that prevents us from working in the future because we might fail again. This fear can stop you from doing better and meeting your goals. Clarence Rivers suggests that instead of letting this fear stop us, we should take the approach of preparing so that we can be successful. This involves doing what is necessary to meet the standard, which includes telling yourself you can do it. Often people get fired up to do something and then talk themselves out of doing it because of negative thoughts and words that they tell themselves. A young boy hears of a footrace at school, and he talks to everyone about how he will enter and win. As the days go by, the young boy starts to think about not winning, and he begins to be afraid of what happens. At first, he shrugs off the thought, but it returns and the more he pays attention to it, the more he talks himself out of the race. He lost before he ever ran. The same thing happens at school and work. At college, many students take the easy classes, so they have better chances of passing. Hard courses require more work, and what if they fail? The answer to that is they do the best they can, and if they fail, they do it again. However, it is easier to rationalize taking the easy courses. A student settles, and instead of becoming an engineer, for example, she majors in business management. Selling herself or himself short has cost her or him a career they wanted.

At work, you get ahead by taking on the tough jobs and completing them right. Unfortunately, you could fail, but the promotions go to the workers that take on the tough challenges and win. Unfortunately, we can talk ourselves out of the tough challenges by telling ourselves that our current job is just fine, the salary is good, or that we are safe where we are. Imagine a young woman who wanted to be a safety manager. She had graduated from college and was working as a secretary. She is offered the chance of being a safety technician. Rather than stay safe, she took on the new job. She suffered for a few months while she learned the new job, but she stuck it out. She later applied for and was accepted into the safety and occupational health intern program with her employer. She kept telling herself she could do it; she had gone to college for a reason, and now she was going to see this through. She graduated and went on to a successful career. Yes, she is now a safety manager.

That is what NLP is. It appears when you hear that inner voice tell you no, when your friends say you would be better off playing it safe, or maybe when your boss tells you that the new job will be pretty hard, and she does not know if you are up to it. Take charge of the situation and reprogram what your inner voice is saying and explain to others that you can do and be more than you are.

The Inner Voice

Clarence Rivers (2013) writes a lot about the inner voice. He describes it as working like our conscience. That means it tells us what to do and how to think and react. Many times, our inner voice is right. For example, we should not take that shortcut to the restaurant through that dark alley. Many times, our inner voice is wrong. For example, I cannot speak to the board of directors about safety without stuttering and stammering. We must analyze what our inner voice is telling us. You have probably experienced an incident where you made an awful decision, like jaywalking. When you are crossing the street, a car turns the corner and swipes you, knocking you down. You are not hurt, but you know you were just lucky. If you are like most of us, you cuss and scream at yourself through your inner voice about how stupid you are, how you could do something so dumb. Then you are upset and feel terrible for hours. You may even continue to tell yourself bad things into the night. The good thing is we can stop this. We can recognize when we do it and reprogram our minds to say it was a bad mistake and that we could have gotten hurt, but that we did not, and we can apply this lesson and not ever jaywalk again.

Personal Standards

Another area that we set ourselves up to fail in is through setting high personal standards that we cannot meet. Clarence Rivers (2013) identifies this as "setting unrealistic standards." The young woman was a safety professional, but she wanted to teach college. That required her to have a master's degree or preferably a Ph.D. in safety. With her current life situation, it was impossible for her to get more education. The time and money needed were out of reach. Unfortunately, she thought about it often and got to the point that she was disappointed in herself for not reaching her dream. Her disappointment began to show at work, and her supervisor counseled her. She continued to be a safety professional, but with her prospects diminished because she focused on a goal that she could not have. We must recognize in life that there are some things we can do and others we simply cannot. If we develop standards that are too high, we cause ourselves to be too critical of our efforts. We may even stop trying because if we cannot have what we want, why bother trying at all? Look at your goals and determine if they are practical. If so, shoot for the stars. If not, identify anther goal that you can reach. I mean for you to set goals that with hard work you can achieve.

Anchors

As human beings, we try to achieve pleasure and avoid pain. We do things we hope gives us something we want, like a paycheck, a diploma, or satisfaction. George Lynch (2014) says that NLP uses our desire to get pleasure and avoid pain to anchor or help us make a permanent change in our behavior. We have probably all heard of a runner who, before a race, sits quietly, thinking through the whole race. She feels the joy and exaltation of the race and the sweet taste of victory. She sees herself winning. She drives out any thought of failure or defeat. When she runs the race, she probably runs it as she saw it in her mind. She is positive and giving her best. Does this mean she wins? No, but it says that she has seen herself as a winner and has given her best at the race. She anchored on the pleasure of running and winning as a source of inspiration. Think about how she would have run if she had thoughts of not being good enough or not fast enough. Chances are she would not have done her best and lost the race. Many athletes swear by this technique and credit it with success. You cannot do this just once and hope to be successful. George Lynch says, "You have to do it daily for a month to make sure that you have installed a suitable and permanent pattern." Let's look at another example. A safety professional will be taking a certification examination. He studies very hard but doubts he can pass. About six weeks before the examination date, he anchors his behavior on the pleasure of being certified, the joy of passing the test, the accolades from his peers. He practices for a month and when he sits for the examination, he is ready. He has a greater chance of passing.

Summary

NLP can work for you personally and professionally. Take stock of your beliefs and values. Are there some changes you want to make? Review the way you set standards. Are you realistic? Listen to what you tell yourself. Are you too harsh or too cynical? I have outlined some important information that I think can help you succeed. Read and understand these examples, identify behaviors you need to change, take care of yourself, like yourself, and believe that you can be successful in your personal and professional lives. SOH managers have a lot on their plate, and the last thing they need is to hurt their chances of success. I encourage you to learn more about NLP. I do not believe all the hype about helping you become rich and famous, but I do believe it works.

CHAPTER 15 - STRESS MANAGEMENT

Introduction

Being a Safety and Occupational Health (SOH) Manager can be stressful. So stressful that it can cause illnesses and damage relationships. Jennifer Green (2014) notes in her book on stress, "Studies have proven that people under stress have an increased vulnerability of contracting a disease and are more susceptible to cardiovascular disease, autoimmune or allergic. In fact, one of the primary symptoms of stress is anxiety. Many have lost their jobs, their homes, their health and sometimes even their sanity." I would be remiss if I did not include a chapter on managing stress to help you balance your life, both to improve your health and give you an opportunity to treat those around you fairly.

I know of several SOH managers with ulcers, myself included. We want to help everyone and at the same time, but we often do not receive all the resources we need to do our job right. This often causes us to work extra hours, stress over the work, and spend time away from family to support our employer's offsite facilities. This results in increased stress.

In this chapter, I will familiarize you with stress and provide some methods to help you control and defuse stress in your work life; however, this will only help if you take the steps to reduce stress in your life.

What are Stress and Anxiety?

The National Institute of Mental Health (NIMH) (2104) defines stress as "the brain's response to any demand." It might surprise you to know that these changes can be positive or negative, and they do not have to be real. The NIMH also says that these changes can be:

- Recurring, short-term, or long-term
- Can be mild and relatively harmless
- Major or extreme

We all experience stress. In fact, stress helps us to perform. Performance gets better as we experience stress; however, at some point the stress fails to improve performance and makes you distressed. At this point, if more stress is added, bad things begin to happen to us. We do not want to eliminate all stress from our lives. We need to find that point in our work life where we excel because of stress and not expose ourselves to any more stress. That is where balancing your life comes in, but more on that later.

Anxiety is a symptom of stress. Kathy Stanton (2014) explains that anxiety makes us feel apprehensive or nervous without cause. She goes on to describe how "the feeling feels abnormal and will adversely affect the life of a person." This is more than just worrying about things; it is worrying constantly. Anxiety can cause more anxiety; it is self-perpetuating. The anxiety can also increase as it builds on the earlier anxiety. The reason I include anxiety in this chapter is because, as Kathy Stanton (2014) says, "stress, if not well managed, will progress into anxiety."

Effects on SOH Managers

All SOH managers will experience stress, and some may experience anxiety. As I mentioned earlier, SOH managers can benefit from stress. When stress affects your work or life in a negative way, it is a problem. Stress can reduce your ability to think and decide. It can cause you to make rash decisions. Unfortunately, SOH managers need to analyze data, which requires effective thinking skills. SOH managers also depend on decision-making skills. Stresses will negatively impact these skills when you need them. Stress can also cause you to suffer from upset stomach, headaches, and rapid heartbeat that can cause you to miss meetings, appointments, and even entire work days.

Some of these symptoms can also cause you to seek medical attention, causing concern from your employer. If stress persists, long term illnesses such as ulcers and other abdominal and intestinal disorders can force you to miss lots of work or even seek another job. Neither of which is good for your career as a SOH manager.

Reducing Stress

Delegation is the first and best way to reduce stress. As a SOH manager, you were likely promoted because you were a really good specialist; however, as a manager, you need to delegate work to the specialists. By shedding work, you can greatly reduce your risk and help those specialists working for you feel better about themselves by showing that you trust them.

Getting enough rest is a great way to reduce stress. Eight hours of sleep is recommended, but it is also important to get those hours during 10:00

pm and 6:00 am. These are the hours your body needs sleep. Often stress can make it hard to go to sleep because you will run the day's events through your mind over and over again. One technique is to say to yourself that you do not want to replay the day's events now, it is time for sleep. I also find it is better to go to bed at the same time each night and get up at the same time each morning. This way your body is ready to sleep.

Breathing is a simple and yet effective way to reduce stress. You can even do it without anyone noticing in the middle of a tense event. Breathing in your abdomen is recommended. You may notice that when you are stressed, you breathe in your chest. This causes you to breathe shallowly and often. Both are bad for you. When you feel stressed, take a deep breath for a count of three. Hold that breath for a count of three. Let that breath out on the count of three. Then hold without a breath for three. Then do it all again for five minutes. You can do this as often as you like. You can also hold for more than three seconds.

Stretching can also reduce the effects of stress. It is best to stretch every day. Nothing fancy is needed; simple stretches are all that is required. Bending over and touching your toes is effective. So is stretching your arms and hands above your head. Hold your head, straight up on your shoulders, then bend your neck and head to the right and hold for a count of five. Move your head back to straight and then bend your neck and head to the left and hold for a count of five. Do this again to the rear and your front. You can do this more than once, and with some practice you can hold for a longer count.

I encourage you to read "The One Minute Manager Gets Fit." This book is part of the One Minute Manager series. It is an easy read and can start you in the right direction. A fitness program does not have to be difficult and rigorous. Something as simple as walking can have great benefit. Exercising can result in the release of natural chemicals in your body to reduce stress.

Eating properly can also help reduce the effects of stress. Reducing your intake of caffeine and other stimulants helps a great deal. Taking time for meals is also important. Stop eating lunch at your desk and go someplace else. Make sure you take the time to eat your meals at home with your family. Instead of sugary snacks, eat fruits and vegetables. Eating a lot of sugar leads to a burst of energy that eventually leads to a sugar slump.

Saying no is another great way to reduce stress. You cannot say no to jobs you are given by your supervisor, but there are a lot of other things you can so no to. I am sure you volunteer, and if so, manage this time well. Doing too much can increase your stress, which can mean you are less effective at work and that could mean you are not doing the great job you want on your main duties. This also applies to activities outside of work. Many people do so many volunteer duties that they cannot do them very well. This can lead to getting a bad reputation. Volunteer only for things you do have time to do.

Summary

In this chapter, I have shared with you the effect stress can have on your work and home life. Stress can negatively affect your work performance or worse. Do not let it get that bad. Take steps to reduce the amount of stress you experience. Reducing stress is an important part of balancing your life. I also gave you some ways that you can reduce stress. The best method is for you to delegate to your direct reports. It takes a load off you and shows you trust them. I also addressed taking better care of yourself physically. Keeping your life in balance will allow you to do a better job at work while strengthening your relationships at work and outside of work. Take steps now to reduce the risks from stress.

SUMMARY

Thank you for choosing this book. I hope you have read the entire book. In these chapters, I have not tried to duplicate the things that you can usually find in traditional management books. What I have attempted to do is focus on the things you probably will not be told or trained on. I have learned these lessons over my career, and they served me well.

I started by addressing the basics of management using behavior rather than just methods. I also believe that building knowledge and managing it properly will provide you with material to make your work consistent and timely. In resource management, I addressed the need for a budget and your role in that process. Now that you are a manager, there are some valuable lessons you need to learn about managing resources. In that same thread, I addressed risk management, which I think is vital for SOH managers to be proficient with.

I used the second thread to address personal skills that I think are essential to your success. I support management over leadership, and think if we spend more time making sure work is done and accurately measured, the better. In this thread, I also highlighted topics that I think you need to be good at to be successful.

In the third thread, I addressed areas that I think are important for you even though many others may not agree with me. There are principles and methods that may not be considered management, but I submit to you are essential to management success.

The last topic I touch on is meant to help you maintain your health and relationships. SOH managers can be exposed to a lot of stress that can damage their health and relationships. Let me close by encouraging you to take care of yourself and those you care for. I and those around you want you to succeed, but we want you to be healthy when you get there.

#####

If you would like to help other readers out, please leave a review of this book on Amazon.com. Your rating and review will help them decide to read or not read this book.

#####

From my other books, I recommend Providing Safe, Healthy and Functional Workspace. You can see it at the following URL Providing Safe, Healthy, and Functional WorkSpaces: A handbook for the New Collateral-Additional Safety Specialist, Fanning, Fred, eBook - Amazon.com

Bibliography

A Guide to the Project Management Body of Knowledge (PMBoK Guide) – Fifth Edition, 2013, USA.

ArcaMax, Classic Quotes by Rosalyn Carter. Retrieved from http://www.arcamax.com/quotes/s-30716-221320 on August 19, 2003.

Aristotle, The Nichomachean Ethics, circa 350-322 BC.

Carnegie, Dale. How to Win Friends and Influence People, revised edition. Simon and Schuster, New York, USA, 1936.

Cashman, Kevin. " Leadership from the Inside Out, 1st Edition" Executive Excellence Publishing, USA, 2001.

Cherniss, Cary. Emotional Intelligence: What it is and Why It Matters, Presented at the Society for Industrial and Organizational Psychology, New Orleans, LA, April 15, 2000.

Clinger, Trevor. A Perfect Environmental and SWOT Analysis Using Saturn Corporation for Example, Tiffin, USA 2014.

Communication Track Toastmasters International (TI). Retrieved from http://www.toastmasters.org/MainMenuCategories/WhatisToastmasters/CommunicationandLeadershipTraining/CommunicationTrack.aspx on August 18, 2009.

Could "personal intelligence" be the future for workplace success? Peter Berry Consultancy. Retrieved from https://www.peterberry.com.au/article/could-personal-intelligence-be-the-future-for-workplace-success on September 21, 2014.

Covey, Stephen. The Seven Habits of Highly Effective People, Free Press, 1989, USA.

Diamond, Lou. Emotional Intelligence, PLR-MMR-Products.com, March 2011, USA.

Education Program Toastmasters International (TI). Retrieved from http://www.toastmasters.org/MainMenuCategories/WhatisToast

masters/CommunicationandLeadershipTraining.aspx on August 18, 2009.

Effective Evaluation, Toastmasters International, 2006, Toastmasters International, Inc., Mission Viejo, USA.

Fanning, Fred. Basic Safety Administration: A Handbook for the New Safety Specialist, Des Plaines, USA, 2003.

Finance and Business, Edition 2013, Version 2.0, International Facility Management Association, Houston, USA 2013.

Fisher, Roger, and Brown, Scott. Getting Together: Building Relationships As We Negotiate, Penguin, 1988, USA.

Fisher, Roger, and Ury, William. Getting to Yes: Negotiating Agreement Without Giving In, Penguin, 1983, USA.

Gautam, Sandeep. The Fundamental Four: Exploring the deepest motivational drives. Retrieved from http://www.psychologytoday.com/essay/the-fundamental-four/201308/personal-intelligence on September 21, 2014.

Goleman, Daniel. Emotional Intelligence 10th Anniversary Edition, 2005, USA.

Green, Jennifer. "How to Eliminate Stress and Anxiety: Beating Stress Step by Step, Ultimate Tips to Manage Stress and Get Rid of Stress Naturally & Easily," Kindle Direct Publishing, September 1, 2014, USA.

Hall, Rayne. SWOT for Writing Success, London, UK, 2015.

Hill, Catie. UNH professor develops personal intelligence theory. Retrieved from http://www.tnhonline.com/news/view.php/159354/UNH-professor-develops-personal-intellig on September 21, 2014.

How Does It Work? Toastmasters International (TI), Retrieved from http://www.toastmasters.org/MainMenuCategories/WhatisToastmasters/HowDoesItWork.aspx on August 18, 2009.

Lambert, David. Body Language 101: The Ultimate Guide to Knowing When People Are Lying, How They Are Feeling, What They are

Thinking, and More. 2007, Skyhorse Publishing. Kindle Edition, London, UK.

Leadership, American Heritage Dictionary. Retrieved from http://dictionary.reference.com/browse/leadership on August 19, 2013.

Lynch George. How to Use NLP: Get Started with Neuro Linguistics Programming Today, BMS Publishing, 2014.

Mayer, John, A.T. Panter, and David Caruso. Does Personal Intelligence Exist? Evidence from a New Ability-Based Measure. Journal of Personality Assessment, Taylor & Francis Group, LLC, 2012.

Mayer, John D. "Know Thyself", Psychology Today, Volume 47, No. 2, April 2014: m pp 65-71.

Mayer, John. The Personality Analyst: How Personality and Personal Intelligence Shape Our Lives. Retrieved from http://www.psychologytoday.com/essay/the-personality-analyst/201404/how-high-is-your-personal-intelligence on September 24, 2014.

McMillan Summaries, Influencer: The Power to Change Anything, Summarized for Busy People, Amazon Kindle Edition, OH, 2013.

Merriam-Webster Dictionary, Influence. Retrieved from http://www.merriam-webster.com/dictionary/influence on May 20, 2014.

Moberg, Anthony. Emotional Intelligence an Introduction, Learning Life Books, 2011, Washington, USA.

Morley, Mike. Emotional Intelligence, PLR-MMR-Products.com, March 2011, USA.

National Institute of Mental Health. "What is Stress?" Retrieved from http://www.nimh.nih.gov/health/publications/stress/index.shtml on December 6, 2014.

Nir, Michael. Influence, and Lead! Fundamentals for Personal and Professional Growth, Kindle Direct Publishing, USA 2013.

Operations and Maintenance, Edition 2013, Version 2.0, International Facility Management Association, Houston, USA 2013.

Ortiz, Margaret. Negotiating Essentials: Theory, Skills, and Practices. Axxa Publishing, 2012, USA.

Personal Intelligence Website Home Page. Retrieved from http://personalintelligence.info/ on September 20, 2014.

PMI Lexicon of Project Management Terms, Project Management Institute, Newtown Square, USA, 2012.

Psychology Today, Psych Basics, First Impressions. Retrieved from http://www.psychologytoday.com/basics/first-impressions on December 28, 2013.

Reiman, Tonya. Building Rapport Through Body Language, 2008-03-20. Retrieved from http://www.tonyareiman.com/articles#tabs-8 on January 12, 2014.

Reiman, Tonya. The Power of Body Language: How to Succeed in Every Business and Social Encounter, 2007, Pocket Books, New York, USA.

Reiman, Tonya. Those First Few Seconds, 2008-01-06. Retrieved from http://www.tonyareiman.com/articles#tabs-19 on January 12, 2014.

Reiman, Tonya. Professional Use of Space, 2008-07-29. Retrieved from http://www.tonyareiman.com/articles#tabs-12 on January 12, 2014.

Rivers, Clarence T. Self-Sabotage: Overcoming Self-Sabotaging Behavior for Life, Kindle Direct Publishing, 2013.

Robinson, Ben. How to Negotiate: Negotiating Tactics That Will Help You to Succeed and Achieve, Kindle Direct Publishing, 2014, USA.

Smedley, Stanley, Ph.D. Personally Speaking, 1966, 1988 Toastmasters International, Inc., Santa Ana, CA, USA.

Stanton, Kathy. "How to Cure Anxiety And Stress: 20 Simple Tips That Lead to A Healthier, Balanced Life," Kindle Direct Publishing, August 14, 2014, USA.

The Art and Science of Leadership, Glossary of Leadership Definitions. Retrieved   from http://www.nwlink.com/~donclark/leader/leaddef.html   on August 19, 2013.

The New Oxford American Dictionary, Oxford University Press, Inc., 2005, USA.

Tideas, Benjamin. Negotiate Anything and Everything: How to Negotiate with Anyone and Win, Kindle Direct Publishing, 2013, USA.

Tool Time Booklet, US Army Management Staff College, Ft Belvoir, VA 2003.

Wyatt, William. Neuro Linguistic Programming Now: How to Achieve Your Goals & Build an Amazing Life through the Power of NLP Techniques, Kindle Direct Publishing, 2013.

ABOUT THE AUTHOR

After a successful career as a Federal Employee that included over twenty years in safety and occupational health. I started writing part-time. My published work includes the peer-reviewed book Basic Safety Administration: A Handbook for the New Safety Specialist in its second edition. I also authored two editions of the peer-reviewed chapter, Safety Training and Documentation Principles published in the bestselling, Safety Professional Handbook, and the Safety Professional Handbook Management Applications, both edited by Joel Haight, Ph.D., CSP. I co-authored the peer-reviewed chapter Safety Training with Christine Fiori, Ph.D., PE, published in the bestselling Construction Safety Management and Engineering, second edition edited by Darryl C. Hill, Ph.D., CSP. The American Society of Safety Professionals Traditionally published my book and chapters.

I self-published another eleven books using Kindle Direct Publishing. Seven of these books are available in paperback and Kindle formats. Four of those books are available only in Kindle format. I have authored over fifty articles in various publications on safety and occupational health and project management. I have earned several writing awards for my non-fiction work and one for my fiction work. I have self-published two novels, A Walk Among the Dead and my most recent Mystery at Devil's Elbow.

I am an Emeritus Professional Member of the American Society of Safety Professionals. I was selected as the Safety Professional of the Year for the Northern Virginia Chapter of this Society. I am also a member of the Non-Fiction Writers Association. I held the Certified Safety Professional (CSP) designation for ten years. I also earned master's degrees from National-Louis University and Webster University.

www.ingramcontent.com/pod-product-compliance
Lightning Source LLC
Chambersburg PA
CBHW071613170526
45166CB00003B/1076